# Corrosion Engineering: An Introduction

# Corrosion Engineering: An Introduction

**Edited by**
Ryan Desario

WILLFORD PRESS
www.willfordpress.com

Published by Willford Press,
118-35 Queens Blvd., Suite 400,
Forest Hills, NY 11375, USA

ISBN: 978-1-68285-483-9

**Cataloging-in-Publication Data**

Corrosion engineering : an introduction / edited by Ryan Desario.
    p. cm.
Includes bibliographical references and index.
ISBN 978-1-68285-483-9
1. Corrosion and anti-corrosives. I. Desario, Ryan.
TA418.74 .C67 2018
620.112 23--dc23

For information on all Willford Press publications
visit our website at www.willfordpress.com

WILLFORD PRESS

# Contents

# Preface

The practice and study of designing, producing and implementing devices, machines, structures, procedures, materials, etc. to control and reduce the effect of corrosion is known as corrosion engineering. The other processes included in this field are fracture, erosion, cracking, crazing, fretting, brittle, etc. The field includes designing some special corrosion prevention processes and methods. This book explores all the important aspects of corrosion engineering in the present day scenario. It discusses in detail about the techniques and theories used in this subject. For someone with an interest and eye for detail, this book covers the most significant topics in the field.

A detailed account of the significant topics covered in this book is provided below:

Chapter 1- One of the processes that occur naturally on the surface of the Earth is corrosion. This process converts metals into a more chemically stable form. Corrosion engineering is a branch of engineering which is devoted to controlling and monitoring corrosion. The chapter on corrosion and corrosion engineering offers an insightful focus, keeping in mind the complex subject matter.

Chapter 2- Corrosion can be of various types. Some of these are rust, galvanic corrosion, crevice corrosion, anaerobic corrosion, corrosion in space, intergranular corrosion, microbial corrosion, laser peening, biogenic sulfide corrosion, metal dusting and pitting corrosion. Corrosion is best understood in confluence with the major topics listed in the following chapter.

Chapter 3- Corrosion is an unwanted effect that occurs in metals. It causes degradation of materials as well as other forms of wear and tear. Tribocorrosion, fretting, stress corrosion cracking are some of the topics explained here. This chapter elucidates the crucial theories and principles of corrosion engineering.

Chapter 4- Conversion coatings are the coatings given to metals that protect them from corrosion. The other ways of preventing corrosion are corrosion inhibitor, cathodic protection, galvanization, coating, low plasticity burnishing, corrosion mapping by ultrasonics, cyclic corrosion testing and salt spray test. The topics discussed in the chapter are of great importance to broaden the existing knowledge on corrosion engineering.

Chapter 5- Plating is a type of surface covering which has been practiced hundreds of years. It is used to improve solderability and to prevent corrosion. Electroplating, chrome plating, etc. are some topics explained in relation to plating. This chapter will provide an integrated understanding of plating.

It gives me an immense pleasure to thank our entire team for their efforts. Finally in the end, I would like to thank my family and colleagues who have been a great source of inspiration and support.

**Editor**

# Understanding Corrosion and Corrosion Engineering

One of the processes that occur naturally on the surface of the Earth is corrosion. This process converts metals into a more chemically stable form. Corrosion engineering is a branch of engineering which is devoted to controlling and monitoring corrosion. The chapter on corrosion and corrosion engineering offers an insightful focus, keeping in mind the complex subject matter.

## Corrosion

Rust, the most familiar example of corrosion

Corrosion is a natural process, which converts a refined metal to a more chemically-stable form, such as its oxide, hydroxide, or sulfide. It is the gradual destruction of materials (usually metals) by chemical and/or electrochemical reaction with their environment. Corrosion engineering is the field dedicated to controlling and stopping corrosion.

In the most common use of the word, this means electrochemical oxidation of metal in reaction with an oxidant such as oxygen or sulfur. Rusting, the formation of iron oxides, is a well-known example of electrochemical corrosion. This type of damage typically produces oxide(s) or salt(s) of the original metal, and results in a distinctive orange colouration. Corrosion can also occur in materials other than metals, such as ceramics or polymers, although in this context, the term "degradation" is more common. Corrosion degrades the useful properties of materials and structures including strength, appearance and permeability to liquids and gases.

Many structural alloys corrode merely from exposure to moisture in air, but the process can be strongly affected by exposure to certain substances. Corrosion can be concentrated locally to form a pit or crack, or it can extend across a wide area more or less uniformly corroding the surface. Because corrosion is a diffusion-controlled process, it occurs on exposed surfaces. As a result, methods to reduce the activity of the exposed surface, such as passivation and chromate conversion, can increase a material's corrosion resistance. However, some corrosion mechanisms are less visible and less predictable.

## Galvanic Corrosion

Galvanic corrosion of aluminium. A 5-mm-thick aluminium alloy plate is physically (and hence, electrically) connected to a 10-mm-thick mild steel structural support. Galvanic corrosion occurred on the aluminium plate along the joint with the steel. Perforation of aluminium plate occurred within 2 years.

Galvanic corrosion occurs when two different metals have physical or electrical contact with each other and are immersed in a common electrolyte, or when the same metal is exposed to electrolyte with different concentrations. In a galvanic couple, the more active metal (the anode) corrodes at an accelerated rate and the more noble metal (the cathode) corrodes at a slower rate. When immersed separately, each metal corrodes at its own rate. What type of metal(s) to use is readily determined by following the galvanic series. For example, zinc is often used as a sacrificial anode for steel structures. Galvanic corrosion is of major interest to the marine industry and also anywhere water (containing salts) contacts pipes or metal structures.

Factors such as relative size of anode, types of metal, and operating conditions (temperature, humidity, salinity, etc.) affect galvanic corrosion. The surface area ratio of the anode and cathode directly affects the corrosion rates of the materials. Galvanic corrosion is often prevented by the use of sacrificial anodes.

## Galvanic Series

In any given environment (one standard medium is aerated, room-temperature seawater), one metal will be either more *noble* or more *active* than others, based on how strongly its ions are bound to the surface. Two metals in electrical contact share the same electrons, so that the "tug-of-war" at each surface is analogous to competition for free electrons between the two materials. Using the electrolyte as a host for the flow of ions in the same direction, the noble metal will take electrons from the active one. The resulting mass flow or electric current can be measured to establish a hierarchy of materials in the medium of interest. This hierarchy is called a *galvanic series* and is useful in predicting and understanding corrosion.

## Corrosion Removal

Often it is possible to chemically remove the products of corrosion. For example, phosphoric acid in the form of naval jelly is often applied to ferrous tools or surfaces to remove rust. Corrosion removal should not be confused with electropolishing, which removes some layers of the underlying metal to make a smooth surface. For example, phosphoric acid may also be used to electropolish copper but it does this by removing copper, not the products of copper corrosion.

## Resistance to Corrosion

Some metals are more intrinsically resistant to corrosion than others. There are various ways of protecting metals from corrosion (oxidation) including painting, hot dip galvanizing, and combinations of these.

## Intrinsic Chemistry

Gold nuggets do not naturally corrode, even on a geological time scale.

The materials most resistant to corrosion are those for which corrosion is thermodynamically unfavorable. Any corrosion products of gold or platinum tend to decompose spontaneously into pure metal, which is why these elements can be found in metallic form on Earth and have long been valued. More common "base" metals can only be protected by more temporary means.

Some metals have naturally slow reaction kinetics, even though their corrosion is thermodynamically favorable. These include such metals as zinc, magnesium, and cadmium. While corrosion of these metals is continuous and ongoing, it happens at an acceptably slow rate. An extreme example is graphite, which releases large amounts of energy upon oxidation, but has such slow kinetics that it is effectively immune to electrochemical corrosion under normal conditions.

## Passivation

Passivation refers to the spontaneous formation of an ultrathin film of corrosion products, known as a passive film, on the metal's surface that act as a barrier to further oxidation. The chemical composition and microstructure of a passive film are different from the underlying metal. Typical passive film thickness on aluminium, stainless steels, and alloys is within 10 nanometers. The passive film is different from oxide layers that are formed upon heating and are in the micrometer thickness range – the passive film recovers if removed or damaged whereas the oxide layer does not. Passivation in natural environments such as air, water and soil at moderate pH is seen in such materials as aluminium, stainless steel, titanium, and silicon.

Passivation is primarily determined by metallurgical and environmental factors. The effect of pH is summarized using Pourbaix diagrams, but many other factors are influential. Some conditions that inhibit passivation include high pH for aluminium and zinc, low pH or the presence of chloride ions for stainless steel, high temperature for titanium (in which case the oxide dissolves into the metal, rather than the electrolyte) and fluoride ions for silicon. On the other hand, unusual conditions may result in passivation of materials that are normally unprotected, as the alkaline environment of concrete does for steel rebar. Exposure to a liquid metal such as mercury or hot solder can often circumvent passivation mechanisms.

## Corrosion in Passivated Materials

Passivation is extremely useful in mitigating corrosion damage, however even a high-quality alloy will corrode if its ability to form a passivating film is hindered. Proper selection of the right grade of material for the specific environment is important for the long-lasting performance of this group of materials. If breakdown occurs in the passive film due to chemical or mechanical factors, the resulting major modes of corrosion may include pitting corrosion, crevice corrosion, and stress corrosion cracking.

## Pitting Corrosion

Certain conditions, such as low concentrations of oxygen or high concentrations of species such as chloride which complete as anions, can interfere with a given alloy's ability to re-form a passivating film. In the worst case, almost all of the surface will remain protected, but tiny local fluctuations will degrade the oxide film in a few critical points. Corrosion at these points will be greatly amplified, and can cause *corrosion pits* of several types, depending upon conditions. While the corrosion pits only nucleate under fairly extreme circumstances, they can continue to grow even when conditions return to normal, since the interior of a pit is naturally deprived of oxygen and locally the pH decreases to very low values and the corrosion rate increases due to an autocatalytic process. In extreme cases, the sharp tips of extremely long and narrow corrosion pits can cause stress concentration to the point that otherwise tough alloys can shatter; a thin film pierced by an invisibly small hole can hide a thumb sized pit from view. These problems are especially dangerous because they are difficult to detect before a part or structure fails. Pitting remains among the most common and damaging forms of corrosion in passivated alloys, but it can be prevented by control of the alloy's environment.

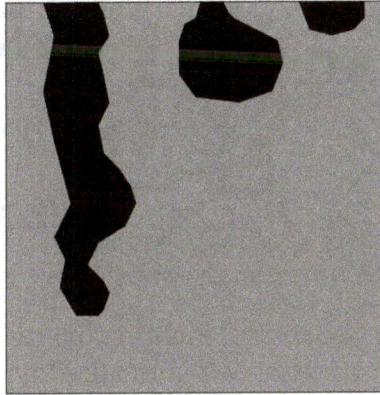

Diagram showing cross-section of pitting corrosion

Pitting results when a small hole, or cavity, forms in the metal, usually as a result of de-passivation of a small area. This area becomes anodic, while part of the remaining metal becomes cathodic, producing a localized galvanic reaction. The deterioration of this small area penetrates the metal and can lead to failure. This form of corrosion is often difficult to detect due to the fact that it is usually relatively small and may be covered and hidden by corrosion-produced compounds.

## Weld Decay and Knifeline Attack

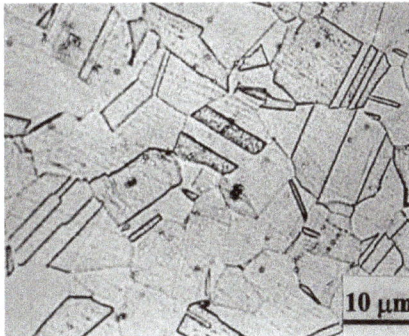

Normal microstructure of Type 304 stainless steel surface

Stainless steel can pose special corrosion challenges, since its passivating behavior relies on the presence of a major alloying component (chromium, at least 11.5%). Because of the elevated temperatures of welding and heat treatment, chromium carbides can form in the grain boundaries of stainless alloys. This chemical reaction robs the material of chromium in the zone near the grain boundary, making those areas much less resistant to corrosion. This creates a galvanic couple with the well-protected alloy nearby, which leads to "weld decay" (corrosion of the grain boundaries in the heat affected zones) in highly corrosive environments. This process can seriously reduce the mechanical strength of welded joints over time.

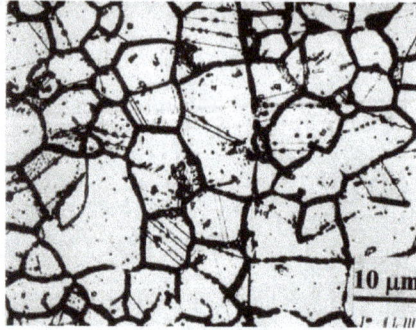

Sensitized metallic microstructure, showing wider intergranular boundaries

A stainless steel is said to be "sensitized" if chromium carbides are formed in the microstructure. A typical microstructure of a normalized type 304 stainless steel shows no signs of sensitization, while a heavily sensitized steel shows the presence of grain boundary precipitates. The dark lines in the sensitized microstructure are networks of chromium carbides formed along the grain boundaries.

Special alloys, either with low carbon content or with added carbon "getters" such as titanium and niobium (in types 321 and 347, respectively), can prevent this effect, but the latter require special heat treatment after welding to prevent the similar phenomenon of "knifeline attack". As its name implies, corrosion is limited to a very narrow zone adjacent to the weld, often only a few micrometers across, making it even less noticeable.

## Crevice Corrosion

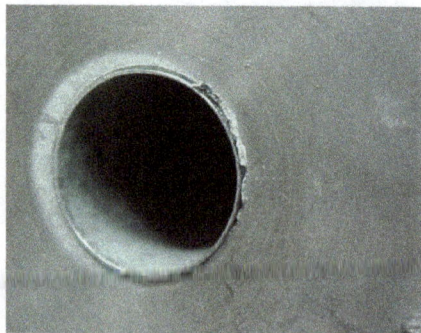

Corrosion in the crevice between the tube and tube sheet
(both made of type 316 stainless steel) of a heat exchanger in a seawater desalination plant

Crevice corrosion is a localized form of corrosion occurring in confined spaces (crevices), to which the access of the working fluid from the environment is limited. Formation of a differential aeration cell leads to corrosion inside the crevices. Examples of crevices are gaps and contact areas between parts, under gaskets or seals, inside cracks and seams, spaces filled with deposits and under sludge piles.

Crevice corrosion is influenced by the crevice type (metal-metal, metal-nonmetal), crevice geometry (size, surface finish), and metallurgical and environmental factors. The susceptibility to crevice corrosion can be evaluated with ASTM standard procedures. A critical crevice corrosion temperature is commonly used to rank a material's resistance to crevice corrosion.

## Microbial Corrosion

Microbial corrosion, or commonly known as microbiologically influenced corrosion (MIC), is a corrosion caused or promoted by microorganisms, usually chemoautotrophs. It can apply to both metallic and non-metallic materials, in the presence or absence of oxygen. Sulfate-reducing bacteria are active in the absence of oxygen (anaerobic); they produce hydrogen sulfide, causing sulfide stress cracking. In the presence of oxygen (aerobic), some bacteria may directly oxidize iron to iron oxides and hydroxides, other bacteria oxidize sulfur and produce sulfuric acid causing biogenic sulfide corrosion. Concentration cells can form in the deposits of corrosion products, leading to localized corrosion.

Accelerated low-water corrosion (ALWC) is a particularly aggressive form of MIC that affects steel piles in seawater near the low water tide mark. It is characterized by an orange sludge, which smells of hydrogen sulfide when treated with acid. Corrosion rates can be very high and design corrosion allowances can soon be exceeded leading to premature failure of the steel pile. Piles that have been coated and have cathodic protection installed at the time of construction are not susceptible to ALWC. For unprotected piles, sacrificial anodes can be installed locally to the affected areas to inhibit the corrosion or a complete retrofitted sacrificial anode system can be installed. Affected areas can also be treated using cathodic protection, using either sacrificial anodes or applying current to an inert anode to produce a calcareous deposit, which will help shield the metal from further attack.

## High-temperature Corrosion

High-temperature corrosion is chemical deterioration of a material (typically a metal) as a result of heating. This non-galvanic form of corrosion can occur when a metal is subjected to a hot atmosphere containing oxygen, sulfur, or other compounds capable of oxidizing (or assisting the oxidation of) the material concerned. For example, materials used in aerospace, power generation and even in car engines have to resist sustained periods at high temperature in which they may be exposed to an atmosphere containing potentially highly corrosive products of combustion.

The products of high-temperature corrosion can potentially be turned to the advantage of the engineer. The formation of oxides on stainless steels, for example, can provide a protective layer preventing further atmospheric attack, allowing for a material to be used for sustained periods at both room and high temperatures in hostile conditions. Such high-temperature corrosion products, in the form of compacted oxide layer glazes, prevent or reduce wear during high-temperature sliding contact of metallic (or metallic and ceramic) surfaces.

## Metal Dusting

Metal dusting is a catastrophic form of corrosion that occurs when susceptible materials are exposed to environments with high carbon activities, such as synthesis gas and other high-CO environments. The corrosion manifests itself as a break-up of bulk metal to metal powder. The suspected mechanism is firstly the deposition of a graphite layer on the surface of the metal, usually from carbon monoxide (CO) in the vapor phase. This graphite layer is then thought to form metastable $M_3C$ species (where M is the metal), which migrate away from the metal surface. However, in some regimes no $M_3C$ species is observed indicating a direct transfer of metal atoms into the graphite layer.

## Protection from Corrosion

The US military shrink wraps equipment such as helicopters to protect
them from corrosion and thus save millions of dollars

Various treatments are used to slow corrosion damage to metallic objects which are exposed to the weather, salt water, acids, or other hostile environments. Some unprotected metallic alloys are extremely vulnerable to corrosion, such as those used in neodymium magnets, which can spall or crumble into powder even in dry, temperature-stable indoor environments unless properly treated to discourage corrosion.

## Surface Treatments

When surface treatments are used to retard corrosion, great care must be taken to ensure complete coverage, without gaps, cracks, or pinhole defects. Small defects can act as an "Achilles' heel", allowing corrosion to penetrate the interior and causing extensive damage even while the outer protective layer remains apparently intact for a period of time.

## Applied Coatings

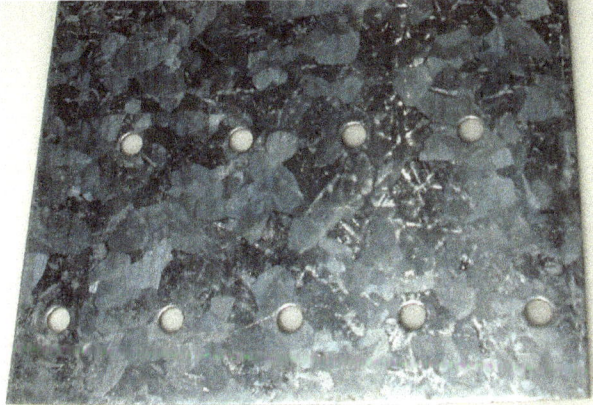

Galvanized surface

Plating, painting, and the application of enamel are the most common anti-corrosion treatments. They work by providing a barrier of corrosion-resistant material between the damaging environment and the structural material. Aside from cosmetic and manufacturing issues, there may be tradeoffs in mechanical flexibility versus resistance to abrasion and high temperature. Platings usually fail only in small sections, but if the plating is more noble than the substrate (for example, chromium on steel), a galvanic couple will cause any exposed area to corrode much more rapidly than an unplated surface would. For this reason, it is often wise to plate with active metal such as zinc or cadmium.

Painting either by roller or brush is more desirable for tight spaces; spray would be better for larger coating areas such as steel decks and waterfront applications. Flexible polyurethane coatings, like Durabak-M26 for example, can provide an anti-corrosive seal with a highly durable slip resistant membrane. Painted coatings are relatively easy to apply and have fast drying times although temperature and humidity may cause dry times to vary.

## Reactive Coatings

If the environment is controlled (especially in recirculating systems), corrosion inhibitors can often be added to it. These chemicals form an electrically insulating or chemically impermeable coating on exposed metal surfaces, to suppress electrochemical reactions. Such methods make the system less sensitive to scratches or defects in the coating, since extra inhibitors can be made available wherever metal becomes exposed. Chemicals that inhibit corrosion include some of the salts in hard water (Roman water systems are famous for their mineral deposits), chromates, phosphates, polyaniline, other conducting polymers and a wide range of specially-designed chemicals that resemble surfactants (i.e. long-chain organic molecules with ionic end groups).

# Anodization

This climbing descender is anodized with a yellow finish.

Aluminium alloys often undergo a surface treatment. Electrochemical conditions in the bath are carefully adjusted so that uniform pores, several nanometers wide, appear in the metal's oxide film. These pores allow the oxide to grow much thicker than passivating conditions would allow. At the end of the treatment, the pores are allowed to seal, forming a harder-than-usual surface layer. If this coating is scratched, normal passivation processes take over to protect the damaged area.

Anodizing is very resilient to weathering and corrosion, so it is commonly used for building facades and other areas where the surface will come into regular contact with the elements. While being resilient, it must be cleaned frequently. If left without cleaning, panel edge staining will naturally occur. Anodization is the process of converting an anode into cathode by bringing a more active anode in contact with it.

# Biofilm Coatings

A new form of protection has been developed by applying certain species of bacterial films to the surface of metals in highly corrosive environments. This process increases the corrosion resistance substantially. Alternatively, antimicrobial-producing biofilms can be used to inhibit mild steel corrosion from sulfate-reducing bacteria.

# Controlled Permeability Formwork

Controlled permeability formwork (CPF) is a method of preventing the corrosion of reinforcement by naturally enhancing the durability of the cover during concrete placement. CPF has been used in environments to combat the effects of carbonation, chlorides, frost and abrasion.

# Cathodic Protection

Cathodic protection (CP) is a technique to control the corrosion of a metal surface by making that surface the cathode of an electrochemical cell. Cathodic protection systems are most commonly used to protect steel, and pipelines and tanks; steel pier piles, ships, and offshore oil platforms.

## Sacrificial Anode Protection

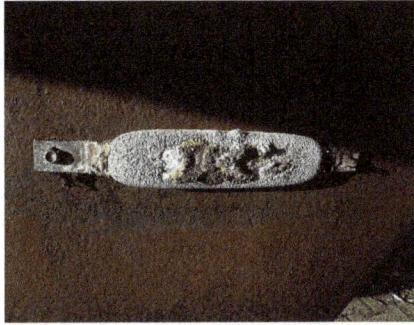

Sacrificial anode attached to the hull of a ship

For effective CP, the potential of the steel surface is polarized (pushed) more negative until the metal surface has a uniform potential. With a uniform potential, the driving force for the corrosion reaction is halted. For galvanic CP systems, the anode material corrodes under the influence of the steel, and eventually it must be replaced. The polarization is caused by the current flow from the anode to the cathode, driven by the difference in electrode potential between the anode and the cathode.

## Impressed Current Cathodic Protection

For larger structures, galvanic anodes cannot economically deliver enough current to provide complete protection. Impressed current cathodic protection (ICCP) systems use anodes connected to a DC power source (such as a cathodic protection rectifier). Anodes for ICCP systems are tubular and solid rod shapes of various specialized materials. These include high silicon cast iron, graphite, mixed metal oxide or platinum coated titanium or niobium coated rod and wires.

## Anodic Protection

Anodic protection impresses anodic current on the structure to be protected (opposite to the cathodic protection). It is appropriate for metals that exhibit passivity (e.g. stainless steel) and suitably small passive current over a wide range of potentials. It is used in aggressive environments, such as solutions of sulfuric acid.

## Rate of Corrosion

A simple test for measuring corrosion is the weight loss method. The method involves exposing a clean weighed piece of the metal or alloy to the corrosive environment for a specified time followed by cleaning to remove corrosion products and weighing the piece to determine the loss of weight. The rate of corrosion (R) is calculated as

$$R = \frac{kW}{\rho At}$$

where $k$ is a constant, $W$ is the weight loss of the metal in time $t$, $A$ is the surface area of the metal exposed, and $\rho$ is the density of the metal (in g/cm³).

Other common expressions for the corrosion rate is penetration depth and change of mechanical properties.

These neodymium magnets corroded extremely rapidly after only 5 months of outside exposure

## Economic Impact

The collapsed Silver Bridge, as seen from the Ohio side

In 2002, the US Federal Highway Administration released a study titled "Corrosion Costs and Preventive Strategies in the United States" on the direct costs associated with metallic corrosion in the US industry. In 1998, the total annual direct cost of corrosion in the U.S. was ca. $276 billion (ca. 3.2% of the US gross domestic product). Broken down into five specific industries, the economic losses are $22.6 billion in infrastructure; $17.6 billion in production and manufacturing; $29.7 billion in transportation; $20.1 billion in government; and $47.9 billion in utilities.

Rust is one of the most common causes of bridge accidents. As rust has a much higher volume than the originating mass of iron, its build-up can also cause failure by forcing apart adjacent parts. It was the cause of the collapse of the Mianus river bridge in 1983, when the bearings rusted internally and pushed one corner of the road slab off its support. Three drivers on the roadway at the time died as the slab fell into the river below.

The following NTSB investigation showed that a drain in the road had been blocked for road re-surfacing, and had not been unblocked; as a result, runoff water penetrated the support hangers. Rust was also an important factor in the Silver Bridge disaster of 1967 in West Virginia, when a steel suspension bridge collapsed within a minute, killing 46 drivers and passengers on the bridge at the time.

Similarly, corrosion of concrete-covered steel and iron can cause the concrete to spall, creating severe structural problems. It is one of the most common failure modes of reinforced concrete bridges. Measuring instruments based on the half-cell potential can detect the potential corrosion spots before total failure of the concrete structure is reached.

Until 20–30 years ago, galvanized steel pipe was used extensively in the potable water systems for single and multi-family residents as well as commercial and public construction. Today, these systems have long ago consumed the protective zinc and are corroding internally resulting in poor water quality and pipe failures. The economic impact on homeowners, condo dwellers, and the public infrastructure is estimated at 22 billion dollars as the insurance industry braces for a wave of claims due to pipe failures.

## Corrosion in Nonmetals

Most ceramic materials are almost entirely immune to corrosion. The strong chemical bonds that hold them together leave very little free chemical energy in the structure; they can be thought of as already corroded. When corrosion does occur, it is almost always a simple dissolution of the material or chemical reaction, rather than an electrochemical process. A common example of corrosion protection in ceramics is the lime added to soda-lime glass to reduce its solubility in water; though it is not nearly as soluble as pure sodium silicate, normal glass does form sub-microscopic flaws when exposed to moisture. Due to its brittleness, such flaws cause a dramatic reduction in the strength of a glass object during its first few hours at room temperature.

## Corrosion of Polymers

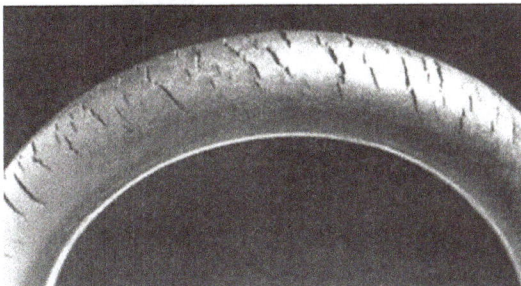

Ozone cracking in natural rubber tubing

Polymer degradation involves several complex and often poorly understood physiochemical processes. These are strikingly different from the other processes discussed here, and so the term "corrosion" is only applied to them in a loose sense of the word. Because of their large molecular weight, very little entropy can be gained by mixing a given mass of polymer with another substance, making them generally quite difficult to dissolve. While dissolution is a problem in some polymer applications, it is relatively simple to design against.

A more common and related problem is "swelling", where small molecules infiltrate the structure, reducing strength and stiffness and causing a volume change. Conversely, many polymers (notably flexible vinyl) are intentionally swelled with plasticizers, which can be leached out of the structure, causing brittleness or other undesirable changes.

The most common form of degradation, however, is a decrease in polymer chain length. Mechanisms which break polymer chains are familiar to biologists because of their effect on DNA: ionizing radiation (most commonly ultraviolet light), free radicals, and oxidizers such as oxygen, ozone, and chlorine. Ozone cracking is a well-known problem affecting natural rubber for example. Plastic additives can slow these process very effectively, and can be as simple as a UV-absorbing pigment (e.g. titanium dioxide or carbon black). Plastic shopping bags often do not include these additives so that they break down more easily as ultrafine particles of litter.

## Corrosion of Glasses

Glass corrosion

Glass is characterized by a high degree of corrosion-resistance. Because of its high water-resistance it is often used as primary packaging material in the pharma industry since most medicines are preserved in a watery solution. Besides its water-resistance, glass is also very robust when being exposed to chemically aggressive liquids or gases. While other materials like metal or plastics quickly reach their limits, special glass-types can easily hold up.

Glass disease is the corrosion of silicate glasses in aqueous solutions. It is governed by two mechanisms: diffusion-controlled leaching (ion exchange) and hydrolytic dissolution of the glass network. Both mechanisms strongly depend on the pH of contacting solution: the rate of ion exchange decreases with pH as $10^{-0.5pH}$ whereas the rate of hydrolytic dissolution increases with pH as $10^{0.5pH}$.

Mathematically, corrosion rates of glasses are characterized by normalized corrosion rates of elements $NR_i$ (g/cm²·d) which are determined as the ratio of total amount of released species into the water $M_i$ (g) to the water-contacting surface area S (cm²), time of contact t (days) and weight fraction content of the element in the glass $f_i$:

$$NR_i = \frac{M_i}{Sf_i t}.$$

The overall corrosion rate is a sum of contributions from both mechanisms (leaching + dissolution) $NR_i = NRx_i + NRh$. Diffusion-controlled leaching (ion exchange) is characteristic of the initial phase of corrosion and involves replacement of alkali ions in the glass by a hydronium ($H_3O^+$) ion from the solution. It causes an ion-selective depletion of near surface layers of glasses and gives an inverse square root dependence of corrosion rate with exposure time. The diffusion-controlled normalized leaching rate of cations from glasses (g/cm²·d) is given by:

$$NRx_i = 2\rho\sqrt{\frac{D_i}{\pi t}},$$

where t is time, $D_i$ is the i-th cation effective diffusion coefficient (cm²/d), which depends on pH of contacting water as $D_i = D_{io} \cdot 10^{-pH}$, and $\rho$ is the density of the glass (g/cm³).

Glass network dissolution is characteristic of the later phases of corrosion and causes a congruent release of ions into the water solution at a time-independent rate in dilute solutions (g/cm²·d):

$$NRh = \rho r_h,$$

where $r_h$ is the stationary hydrolysis (dissolution) rate of the glass (cm/d). In closed systems the consumption of protons from the aqueous phase increases the pH and causes a fast transition to hydrolysis. However, a further saturation of solution with silica impedes hydrolysis and causes the glass to return to an ion-exchange, e.g. diffusion-controlled regime of corrosion.

In typical natural conditions normalized corrosion rates of silicate glasses are very low and are of the order of $10^{-7}$–$10^{-5}$ g/(cm²·d). The very high durability of silicate glasses in water makes them suitable for hazardous and nuclear waste immobilisation.

# Glass Corrosion Tests

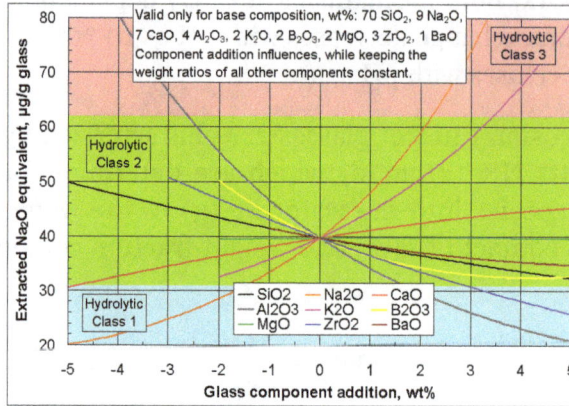

Effect of addition of a certain glass component on the chemical
durability against water corrosion of a specific base glass (corrosion test ISO 719).

There exist numerous standardized procedures for measuring the corrosion (also
called chemical durability) of glasses in neutral, basic, and acidic environments, under
simulated environmental conditions, in simulated body fluid, at high temperature and
pressure, and under other conditions.

The standard procedure ISO 719 describes a test of the extraction of water-soluble basic
compounds under neutral conditions: 2 g of glass, particle size 300–500 µm, is kept for
60 min in 50 ml de-ionized water of grade 2 at 98 °C; 25 ml of the obtained solution is
titrated against 0.01 mol/l HCl solution. The volume of HCl required for neutralization
is classified according to the table below.

| Amount of 0.01M HCl needed to neutralize extracted basic oxides, ml | Extracted $Na_2O$ equivalent, µg | Hydrolytic class |
|---|---|---|
| < 0.1 | < 31 | 1 |
| 0.1-0.2 | 31-62 | 2 |
| 0.2-0.85 | 62-264 | 3 |
| 0.85-2.0 | 264-620 | 4 |
| 2.0-3.5 | 620-1085 | 5 |
| > 3.5 | > 1085 | > 5 |

The standardized test ISO 719 is not suitable for glasses with poor or not extractable
alkaline components, but which are still attacked by water, e.g. quartz glass, $B_2O_3$ glass
or $P_2O_5$ glass.

Usual glasses are differentiated into the following classes:

Hydrolytic class 1 (Type I):

This class, which is also called neutral glass, includes borosilicate glasses (e.g. Duran,
Pyrex, Fiolax).

Glass of this class contains essential quantities of boron oxides, aluminium oxides and alkaline earth oxides. Through its composition neutral glass has a high resistance against temperature shocks and the highest hydrolytic resistance. Against acid and neutral solutions it shows high chemical resistance, because of its poor alkali content against alkaline solutions.

Hydrolytic class 2 (Type II):

This class usually contains sodium silicate glasses with a high hydrolytic resistance through surface finishing. Sodium silicate glass is a silicate glass, which contains alkali- and alkaline earth oxide and primarily sodium oxide and Calcium oxide.

Hydrolytic class 3 (Type III):

Glass of the 3rd hydrolytic class usually contains sodium silicate glasses and has a mean hydrolytic resistance, which is two times poorer than of type 1 glasses.

Acid class DIN 12116 and alkali class DIN 52322 (ISO 695) are to be distinguished from the hydrolytic class DIN 12111 (ISO 719).

## Internal Oxidation

Internal oxidation, in corrosion of metals, is the process of formation of corrosion products (e.g. a metal oxide) within the metal bulk. In other words, the corrosion products are created away from the metal surface, and they are isolated from the surface.

Internal oxidation occurs when some components of the alloy are oxidized in preference to the balance of the bulk. The oxidizer is often oxygen diffusing through the metal bulk from the interface, but it can be also another element (for example sulfur or nitrogen).

Internal oxidation is a well-known corrosion mechanism of nickel-based alloys in the temperature range of 500 to 1200 °C.

Internal oxidation is distinct from selective leaching.

# Corrosion Engineering

Corrosion Engineering is the specialist discipline of applying scientific knowledge, natural laws and physical resources in order to design and implement materials, structures, devices, systems and procedures to manage the natural phenomenon known as corrosion. Generally related to Metallurgy, Corrosion Engineering also relates to non-metallics including ceramics. Corrosion Engineers often manage other not-strictly-corrosion processes including (but not restricted to) cracking, brittle fracture, crazing, fretting, erosion and more.

In the year 1995, it was reported that the costs nationwide in the U.S of corrosion were nearly $300 billion per year.

Corrosion engineering groups have formed around the world in order to prevent, slow and manage the effects of corrosion. Examples of such groups are the National Association of Corrosion Engineers (NACE) and the European Federation of Corrosion (EFC). The corrosion engineers main task is to economically and safely manage the effects of corrosion on materials. Corrosion Engineering master's degree courses are available worldwide and are concerned with the control and understanding of corrosion.

Zaki Ahmad in his book "Principles of corrosion engineering and corrosion control"(10) states that "Corrosion engineering is the application of the principles evolved from corrosion science to minimize or prevent corrosion. Corrosion engineering involves designing of corrosion prevention schemes and implementation of specific codes and practices. Corrosion prevention measures, like cathodic protection, designing to prevent corrosion and coating of structures fall within the regime of corrosion engineering. However, corrosion science and engineering go hand-in-hand and they cannot be separated: it is a permanent marriage to produce new and better methods of protection from time to time". In the "Handbook of corrosion engineering" (4) the author Pierre R. Roberge states "Corrosion is the destructive attack of a material by reaction with its environment. The serious consequences of the corrosion process have become a problem of worldwide significance".

# Types of Corrosion

Corrosion can be of various types. Some of these are rust, galvanic corrosion, crevice corrosion, anaerobic corrosion, corrosion in space, intergranular corrosion, microbial corrosion, laser peening, biogenic sulfide corrosion, metal dusting and pitting corrosion. Corrosion is best understood in confluence with the major topics listed in the following chapter.

## Rust

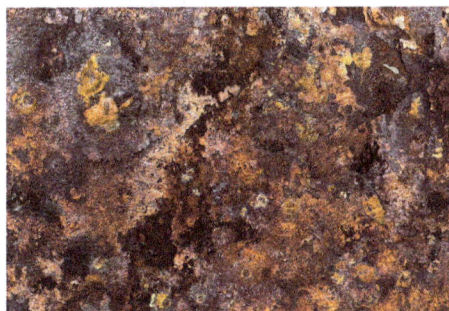

Colors and porous surface texture of rust

Rust is an iron oxide, usually red oxide formed by the redox reaction of iron and oxygen in the presence of water or air moisture. Several forms of rust are distinguishable both visually and by spectroscopy, and form under different circumstances. Rust consists of hydrated iron(III) oxides $Fe_2O_3 \cdot nH_2O$ and iron(III) oxide-hydroxide (FeO(OH), $Fe(OH)_3$).

Given sufficient time, oxygen, and water, any iron mass will eventually convert entirely to rust and disintegrate. Surface rust is flaky and friable, and it provides no protection to the underlying iron, unlike the formation of patina on copper surfaces. Rusting is the common term for corrosion of iron and its alloys, such as steel. Many other metals undergo similar corrosion, but the resulting oxides are not commonly called rust.

Other forms of rust exist, like the result of reactions between iron and chloride in an environment deprived of oxygen. Rebar used in underwater concrete pillars, which generates green rust, is an example. Although rusting is generally a negative aspect of iron, a particular form of rusting, known as "stable rust," causes the object to have a thin coating of rust over the top, and if kept in low relative humidity, makes the "stable" layer protective to the iron below, but not to the extent of other oxides, such as aluminum.

# Chemical Reactions

Heavy rust on the links of a chain near the Golden Gate Bridge in San Francisco; it was continuously exposed to moisture and salt spray, causing surface breakdown, cracking, and flaking of the metal.

Outdoor *Rust Wedge* display at the Exploratorium shows the enormous expansive force of rusting iron

Rust is another name for iron oxide, which occurs when iron or an alloy that contains iron, like steel, is exposed to oxygen and moisture for a long period of time. Over time, the oxygen combines with the metal at an atomic level, forming a new compound called an oxide and weakening the bonds of the metal itself. Although some people refer to rust generally as "oxidation", that term is much more general; although rust forms when iron undergoes oxidation, not all oxidation forms rust. Only iron or alloys that contain iron can rust, but other metals can corrode in similar ways.

The main catalyst for the rusting process is water. Iron or steel structures might appear to be solid, but water molecules can penetrate the microscopic pits and cracks in any exposed metal. The hydrogen atoms present in water molecules can combine with other elements to form acids, which will eventually cause more metal to be exposed. If chloride ions are present, as is the case with saltwater, the corrosion is likely to occur more quickly. Meanwhile, the oxygen atoms combine with metallic atoms to form the destructive oxide compound. As the atoms combine, they weaken the metal, making the structure brittle and crumbly.

## Oxidation of Iron

When impure (cast) iron is in contact with water, oxygen, other strong oxidants, or acids, it rusts. If salt is present, for example in seawater or salt spray, the iron tends to rust more quickly, as a result of electrochemical reactions. Iron metal is relatively unaffected by pure water or by dry oxygen. As with other metals, like aluminium, a tightly adhering oxide coating, a passivation layer, protects the bulk iron from further oxidation. The conversion of the passivating ferrous oxide layer to rust results from the combined action of two agents, usually oxygen and water.

Other degrading solutions are sulfur dioxide in water and carbon dioxide in water. Under these corrosive conditions, iron hydroxide species are formed. Unlike ferrous oxides, the hydroxides do not adhere to the bulk metal. As they form and flake off from the surface, fresh iron is exposed, and the corrosion process continues until either all of the iron is consumed or all of the oxygen, water, carbon dioxide, or sulfur dioxide in the system are removed or consumed.

When iron rusts, the oxides take up more volume than the original metal; this expansion can generate enormous forces, damaging structures made with iron.

## Associated Reactions

The rusting of iron is an electrochemical process that begins with the transfer of electrons from iron to oxygen. The iron is the reducing agent (gives up electrons) while the oxygen is the oxidising agent (gains electrons). The rate of corrosion is affected by water and accelerated by electrolytes, as illustrated by the effects of road salt on the corrosion of automobiles. The key reaction is the reduction of oxygen:

$$O_2 + 4e^- + 2H_2O \rightarrow 4OH^-$$

Because it forms hydroxide ions, this process is strongly affected by the presence of acid. Indeed, the corrosion of most metals by oxygen is accelerated at low pH. Providing the electrons for the above reaction is the oxidation of iron that may be described as follows:

$$Fe \rightarrow Fe^{2+} + 2e^-$$

The following redox reaction also occurs in the presence of water and is crucial to the formation of rust:

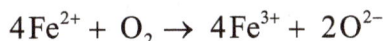

$$4Fe^{2+} + O_2 \rightarrow 4Fe^{3+} + 2O^{2-}$$

In addition, the following multistep acid–base reactions affect the course of rust formation:

$$Fe^{2+} + 2H_2O \rightleftharpoons Fe(OH)_2 + 2H^+$$
$$Fe^{3+} + 3H_2O \rightleftharpoons Fe(OH)_3 + 3H^+$$

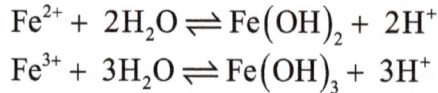

as do the following dehydration equilibria:

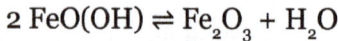

$$Fe(OH)_2 \rightleftharpoons FeO + H_2O$$

$$Fe(OH)_3 \rightleftharpoons FeO(OH) + H_2O$$

$$2\,FeO(OH) \rightleftharpoons Fe_2O_3 + H_2O$$

From the above equations, it is also seen that the corrosion products are dictated by the availability of water and oxygen. With limited dissolved oxygen, iron(II)-containing materials are favoured, including FeO and black lodestone or magnetite ($Fe_3O_4$). High oxygen concentrations favour ferric materials with the nominal formulae $Fe(OH)_{3-x}O_{x/2}$. The nature of rust changes with time, reflecting the slow rates of the reactions of solids.

Furthermore, these complex processes are affected by the presence of other ions, such as $Ca^{2+}$, which serve as electrolytes which accelerate rust formation, or combine with the hydroxides and oxides of iron to precipitate a variety of Ca, Fe, O, OH species.

Onset of rusting can also be detected in laboratory with the use of ferroxyl indicator solution. The solution detects both $Fe^{2+}$ ions and hydroxyl ions. Formation of ions and hydroxyl ions are indicated by blue and pink patches respectively.

## Prevention

Cor-Ten is a special iron alloy that rusts, but still retains its structural integrity

Because of the widespread use and importance of iron and steel products, the prevention or slowing of rust is the basis of major economic activities in a number of specialized technologies.

Rust is permeable to air and water, therefore the interior metallic iron beneath a rust layer continues to corrode. Rust prevention thus requires coatings that preclude rust formation.

## Rust-resistant Alloys

Stainless steel forms a passivation layer of chromium(III) oxide. Similar passivation behavior occurs with magnesium, titanium, zinc, zinc oxides, aluminium, polyaniline, and other electroactive conductive polymers.

Special "weathering steel" alloys such as Cor-Ten rust at a much slower rate than normal, because the rust adheres to the surface of the metal in a protective layer. Designs using this material must include measures that avoid worst-case exposures, since the material still continues to rust slowly even under near-ideal conditions.

## Galvanization

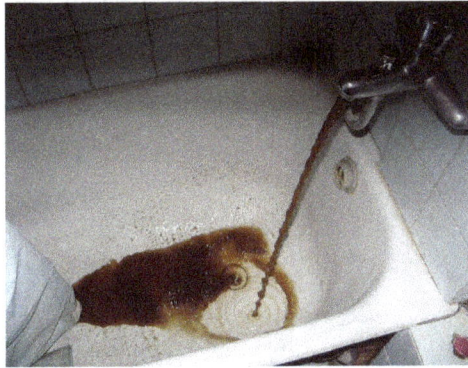

Interior rust in old galvanized iron water pipes can result in brown and black water.

Galvanization consists of an application on the object to be protected of a layer of metallic zinc by either hot-dip galvanizing or electroplating. Zinc is traditionally used because it is cheap, adheres well to steel, and provides cathodic protection to the steel surface in case of damage of the zinc layer. In more corrosive environments (such as salt water), cadmium plating is preferred. Galvanization often fails at seams, holes, and joints where there are gaps in the coating. In these cases, the coating still provides some partial cathodic protection to iron, by acting as a galvanic anode and corroding itself instead of the underlying protected metal. The protective zinc layer is consumed by this action, and thus galvanization provides protection only for a limited period of time.

More modern coatings add aluminium to the coating as *zinc-alume*; aluminium will migrate to cover scratches and thus provide protection for a longer period. These approaches rely on the aluminium and zinc oxides reprotecting a once-scratched surface, rather than oxidizing as a sacrificial anode as in traditional galvanized coatings. In some cases, such as very aggressive environments or long design life, both zinc and a coating are applied to provide enhanced corrosion protection.

Typical galvanization of steel products which are to subject to normal day to day weathering in an outside environment consists of a hot dipped 85 μm zinc coating. Under normal weather conditions, this will deteriorate at a rate of 1 μm per year, giving approximately 85 years of protection.

## Cathodic Protection

Cathodic protection is a technique used to inhibit corrosion on buried or immersed structures by supplying an electrical charge that suppresses the electrochemical reaction. If correctly applied, corrosion can be stopped completely. In its simplest form, it is achieved by attaching a sacrificial anode, thereby making the iron or steel the cathode in the cell formed. The sacrificial anode must be made from something with a more negative electrode potential than the iron or steel, commonly zinc, aluminium, or magnesium. The sacrificial anode will eventually corrode away, ceasing its protective action unless it is replaced in a timely manner.

Cathodic protection can also be provided by using a special-purpose electrical device to appropriately induce an electric charge.

## Coatings and Painting

Flaking paint, exposing a patch of surface rust on sheet metal

Rust formation can be controlled with coatings, such as paint, lacquer, or varnish that isolate the iron from the environment. Large structures with enclosed box sections, such as ships and modern automobiles, often have a wax-based product (technically a "slushing oil") injected into these sections. Such treatments usually also contain rust inhibitors. Covering steel with concrete can provide some protection to steel because of the alkaline pH environment at the steel–concrete interface. However rusting of steel in concrete can still be a problem, as expanding rust can fracture or slowly "explode" concrete from within.

As a closely related example, iron bars were used to reinforce stonework of the Parthenon in Athens, Greece, but caused extensive damage by rusting, swelling, and shattering the marble components of the building.

When only temporary protection is needed for storage or transport, a thin layer of oil, grease, or a special mixture such as Cosmoline can be applied to an iron surface. Such treatments are extensively used when "mothballing" a steel ship, automobile, or other equipment for long-term storage.

Special antiseize lubricant mixtures are available, and are applied to metallic threads and other precision machined surfaces to protect them from rust. These compounds usually contain grease mixed with copper, zinc, or aluminium powder, and other proprietary ingredients.

## Bluing

Bluing is a technique that can provide limited resistance to rusting for small steel items, such as firearms; for it to be successful, a water-displacing oil is rubbed onto the blued steel and other steel.

## Inhibitors

Corrosion inhibitors, such as gas-phase or volatile inhibitors, can be used to prevent corrosion inside sealed systems. They are not effective when air circulation disperses them, and brings in fresh oxygen and moisture.

## Humidity Control

Rust can be avoided by controlling the moisture in the atmosphere. An example of this is the use of silica gel packets to control humidity in equipment shipped by sea.

## Treatment

Rust removal from small iron or steel objects by electrolysis can be done in a home workshop using simple materials such as a plastic bucket, tap water, lengths of rebar, washing soda, baling wire, and a battery charger.

Rust may be treated with commercial products known as rust converter which contain tannic acid which combines with rust.

## Economic Effect

Rust is associated with degradation of iron-based tools and structures. As rust has a much higher volume than the originating mass of iron, its buildup can also cause failure by forcing apart adjacent parts — a phenomenon sometimes known as "rust packing". It was the cause of the collapse of the Mianus river bridge in 1983, when the bearings rusted internally and pushed one corner of the road slab off its support.

Rusting rebar has expanded and spalled concrete off the surface of this reinforced concrete support

Rust was an important factor in the Silver Bridge disaster of 1967 in West Virginia, when a steel suspension bridge collapsed in less than a minute, killing 46 drivers and passengers on the bridge at the time. The Kinzua Bridge in Pennsylvania was blown down by a tornado in 2003, largely because the central base bolts holding the structure to the ground had rusted away, leaving the bridge anchored by gravity alone.

Reinforced concrete is also vulnerable to rust damage. Internal pressure caused by expanding corrosion of concrete-covered steel and iron can cause the concrete to spall, creating severe structural problems. It is one of the most common failure modes of reinforced concrete bridges and buildings.

- Structural failures caused by rust

The Kinzua Bridge After It Collapsed

## Cultural Symbolism

Rust is a commonly used metaphor for slow decay due to neglect, since it gradually converts robust iron and steel metal into a soft crumbling powder. A wide section of the industrialized American Midwest and American Northeast, once dominated by steel foundries, the automotive industry, and other manufacturers, has experienced harsh economic cutbacks that have caused the region to be dubbed the "Rust Belt".

In music, literature, and art, rust is associated with images of faded glory, neglect, decay, and ruin.

## Green Rust

Green rust is a generic name for various green crystalline chemical compounds containing iron(II) and iron(III) cations, the hydroxide ($HO^-$) anion, and another anion such as carbonate ($CO_3^{2-}$), chloride ($Cl^-$), or sulfate ($SO_4^{2-}$), in a layered double hydroxide structure. The most studied varieties are

- carbonate green rust - GR($CO_3^{2-}$): $[Fe_4^{2}Fe_2^{3+}(HO^-)_{12}]^{2+} \cdot [CO_3^{2-} \cdot 2H_2O]^{2-}$.

- chloride green rust - GR($Cl^-$): $[Fe_3^{2+}Fe^{3+}(HO^-)_8]^+ \cdot [Cl^- \cdot nH_2O]^-$.

- sulfate green rust - GR($SO_4^{2-}$): $[Fe_4^{2+}Fe_2^{3+}(HO^-)_{12}]^{2+} \cdot [SO_4^{2-} \cdot 2H_2O]^{2-}$.

Other varieties reported in the literature are bromide $Br^-$, fluoride $F^-$, iodide $I^-$, nitrate $NO_3^-$, and selenate.

Green rust was first recognized as a corrosion crust on iron and steel surfaces. It occurs in nature as the mineral fougerite.

## Structure

The crystal structure of green rust can be understood as the result of inserting the foreign anions and water molecules between brucite-like layers of iron(II) hydroxide, $Fe(OH)_2$. The latter has an hexagonal structure, with layer sequence AcBAcB..., where A and B are planes of hydroxide ions, and c those of $Fe^{2+}$ (iron(II), ferrous) cations. In the green rust, some $Fe^{2+}$ 4cations get oxidized to $Fe^{3+}$ (iron(III), ferric). Each triple layer AcB, which is electrically neutral in the hydroxide, becomes positively charged. The anions then intercalate between those triple layers and restore neutrality.

There are two basic structures of green rust, "type 1" and "type 2". Type 1 is exemplified by the chloride and carbonate varieties. It has a rhombohedral cystal structure similar to that of pyroaurite. The layers are stacked in the sequence AcBiBaCjCbAkA ...; where A, B, and C represent $HO^-$ planes, a, b, and c are layers of mixed $Fe^{2+}$ and $Fe^{3+}$ cations, and i, j, and k are layers of the intercalated anions and water molecules. The c crystallographic parameter is 22.5-22.8 Å for the carbonate, and about 24 Å for the chloride.

Type 2 green rust is exemplified by the sulfate variety. It has a hexagonal crystal structure, with layers probably stacked in the sequence AcBiAbCjA...

## Chemical Properties

In oxidizing environment, green rust generally turns into $Fe^{3+}$ oxyhydroxides, namely α-FeOOH (goethite) and γ-FeOOH (lepidocrocite).

Oxidation of the carbonate variety can be retarded by wetting the material with hydroxyl-containing compounds such as glicerol or glucose, even though they do not penetrate the structure. Some variety of green rust is stabilized also by an atmosphere with high $CO_2$ partial pressure.

Sulfate green rust has been shown to reduce nitrate $NO_3^-$ and nitrite $NO_2^-$ in solution to ammonium $NO_4^+$, with concurrent oxidation of $Fe^{2+}$ to $Fe^{3+}$. Depending on the cations in the solution, the nitrate anions replaced the sulfate in the intercalation layer, before the reduction. It was conjectured that gren rust may be formed in the reducing alkaline conditions below the surface of marine sediments, and may be connected to the disappearance of oxidized species like nitrate in that environment.

Suspensions of carbonate green rust and orange $\gamma$-FeOOH in water will react over a few days produce a black precipitate of magnetite $Fe_3O_4$.

## Occurrence

## Iron and Steel Corrosion

Green rust compounds were identified in green corrosion crusts that form on iron and steel surfaces, in alternating aerobic and anaerobic conditions, by water containing anions such as chloride, sulfate, carbonate, or bicarbonate. They are believed to be intermediates in the oxidative corrosion of iron to form iron(III) oxyhydroxides (ordinary brown rust). The green rust may be formed either directly from metallic iron or from iron(II) hydroxide $Fe(OH)_2$.

## Soil

On the basis of Mössbauer spectroscopic analysis, green rust minerals are suspected to occur as minerals in certain bluish-green soils that are formed in alternating redox conditions, and turn ochre once exposed to air. The green rust has been conjectured to be present in the forn of the mineral fougerite.

## Biologically Mediated Formation

Hexagonal cystals of green rust (carbonate and/or sulfate) have also been obtained as a byproducts of bioreduction of ferric oxyhydroxides by dissimilatory iron-reducing bacteria, such as *Shewanella putrefaciens*, that couple the reduction of $Fe^{3+}$ with the oxidation of organic matter. This process has been conjectured to occur in soil solutions and aquifers.

In one experiment, a 160 mM suspension of orange lepidocrocite $\gamma$-FeOOH in a solution containing formate ($HC_2^-$), incubated for 3 days with a culture of *S. putrefaciens*, turned dark green due to the conversion of the hydroxide to $GR(CO_3^{2-})$, in the form of hexagonal platelets with diameter ~7 μm. In this process, the formate was oxidized to bicarbonate $HC_3^-$ which provided the carbonate anions for the formation of the green rust. The live bacteria were shown to be necessary for the formation of the green rust.

## Laboratory Preparation

### Air Oxidation Methods

Green rust compounds can be synthesized at ordinary ambient temperature and pressure, from solutions containing iron(II) cations, hydroxide anions, and the appropriate intercalatory anions, such as chloride, sulfate, or carbonate.

The result is a suspension of ferrous hydroxide $Fe(OH)_2$ in a solution of the third anion. This suspension is oxidized by stirring in air, or bubbling air through it. Since the product is very prone to oxidation, it is necessary to monitor the process and exclude oxygen once the desired ratio of $Fe^{2+}$ and $Fe^{3+}$ is achieved.

One method first combines an iron(II) salt with sodium hydroxide NaOH to form the ferrous hydroxide suspension. Then the sodium salt of the third anion is added, and the suspension is oxidized by stirring in air.

For example, carbonate green rust can be prepared by mixing solutions of iron(II) sulfate $FeSO_4$ and sodium hydroxide; then adding sufficient amount of sodium carbonate $Na_2CO_3$ solution, followed by the air oxidation step.

Sulfate green rust can be obtained by mixing solutions of $FeCl_2 \cdot 4H_2O$ and NaOH to precipitate $Fe(OH)_2$ then immediately adding sodium sulfate $Na_2SO_4$ and proceeding to the air oxidation step.

A more direct method combines a solution of iron(II) sulfate $FeSO_4$ with NaOH, and proceeding to the oxidizing step. The suspension must have a slight excess of $FeSO_4$ (in the ratio of 0.5833 Fe2+ for each HO– ) for the green rust to form; however, too much of it will produce instead an insoluble basic iron sulfate, iron(II) sulfate hydroxide $Fe_2SO_4(OH)_2 \cdot nH_2O$. The production of green rust is reduced as temperature increases.

### Stoichometric Fe(II) - Fe(III) Methods

An alternate preparation of carbonate green rust first creates a suspension of iron(III) hydroxide $Fe(OH)_3$ in an iron(II) chloride $FeCl_2$ solution, and bubbles carbon dioxide through it.

In a more recent variant, solutions of both iron(II) and iron(III) salts are first mixed, then a solution of NaOH is added, all in the stoichometric proportions of the desired green rust. No oxidation step is then necessary.

## Electrochemistry

Carbonate green rust films have also been obtained from the electrochemical oxidation of iron plates.

# Galvanic Corrosion

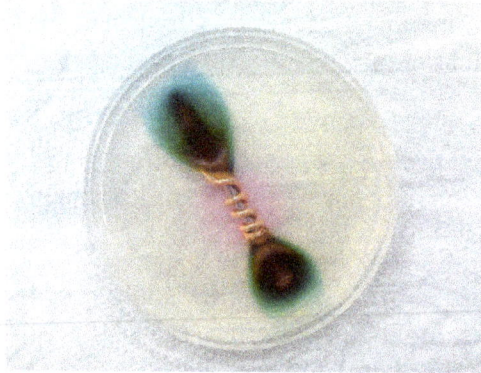

Corrosion of an iron nail wrapped in bright copper wire, showing cathodic protection of copper;
a ferroxyl indicator solution shows colored chemical indications of two
types of ions diffusing through a moist agar medium.

Galvanic corrosion (also called bimetallic corrosion) is an electrochemical process in which one metal corrodes preferentially when it is in electrical contact with another, in the presence of an electrolyte. A similar galvanic reaction is exploited in primary cells to generate a useful electrical voltage to power portable devices.

## Overview

Dissimilar metals and alloys have different electrode potentials, and when two or more come into contact in an electrolyte, one metal acts as anode and the other as cathode. If the electrolyte contains only metal ions that are not easily reduced (such as $Na^+$, $Ca^{2+}$, $K^+$, $Mg^{2+}$, or $Zn^{2+}$), the cathode reaction is reduction of dissolved $H^+$ to $H_2$ or $O_2$ to $OH^-$. The electropotential difference between the reactions at the two electrodes is the driving force for an accelerated attack on the anode metal, which dissolves into the electrolyte. This leads to the metal at the anode corroding more quickly than it otherwise would and corrosion at the cathode being inhibited. The presence of an electrolyte and an electrical conducting path between the metals is essential for galvanic corrosion to occur. The electrolyte provides a means for ion migration whereby ions move to prevent charge build-up that would otherwise stop the reaction.

In some cases, this type of reaction is intentionally encouraged. For example, low-cost household batteries typically contain carbon-zinc cells. As part of a closed circuit (the electron pathway), the zinc within the cell will corrode preferentially (the ion pathway) as an essential part of the battery producing electricity. Another example is the cathodic protection of buried or submerged structures as well as hot water storage tanks. In this case, sacrificial anodes work as part of a galvanic couple, promoting corrosion of the anode, while protecting the cathode metal.

In other cases, such as mixed metals in piping (for example, copper, cast iron and other cast metals), galvanic corrosion will contribute to accelerated corrosion of parts of the system. Corrosion inhibitors such as sodium nitrite or sodium molybdate can be injected into these systems to reduce the galvanic potential. However, the application of these corrosion inhibitors must be monitored closely. If the application of corrosion inhibitors increases the conductivity of the water within the system, the galvanic corrosion potential can be greatly increased.

Acidity or alkalinity (pH) is also a major consideration with regard to closed loop bimetallic circulating systems. Should the pH and corrosion inhibition doses be incorrect, galvanic corrosion will be accelerated. In most HVAC systems, the use of sacrificial anodes and cathodes is not an option, as they would need to be applied within the plumbing of the system and, over time, would corrode and release particles that could cause potential mechanical damage to circulating pumps, heat exchangers, etc.

## Examples of Corrosion

A common example of galvanic corrosion occurs in corrugated iron, a sheet of iron or steel covered with a zinc coating. Even when the protective zinc coating is broken, the underlying steel is not attacked. Instead, the zinc is corroded because it is less noble; only after it has been consumed can rusting of the base metal occur in earnest. By contrast, with a traditional tin can, the opposite of a protective effect occurs: because the tin is more noble than the underlying steel, when the tin coating is broken, the steel beneath is immediately attacked preferentially.

## Statue of Liberty

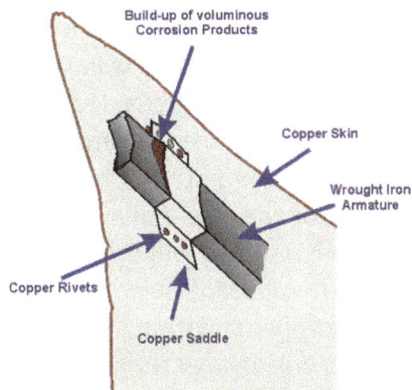

Galvanic corrosion in the Statue of Liberty

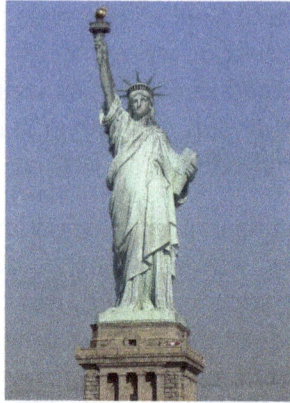

Regular maintenance checks discovered that the Statue
of Liberty suffered from galvanic corrosion.

A spectacular example of galvanic corrosion occurred in the Statue of Liberty when regular maintenance checks in the 1980s revealed that corrosion had taken place between the outer copper skin and the wrought iron support structure. Although the problem had been anticipated when the structure was built by Gustave Eiffel to Frédéric Bartholdi's design in the 1880s, the insulation layer of shellac between the two metals had failed over time and resulted in rusting of the iron supports. An extensive renovation requiring complete disassembly of the statue replaced the original insulation with PTFE. The structure was far from unsafe owing to the large number of unaffected connections, but it was regarded as a precautionary measure to preserve a national symbol of the United States.

## Royal Navy and HMS Alarm

In 17th-century England, Samuel Pepys (then serving as Admiralty Secretary), agreed to the removal of lead sheathing from British Royal Navy vessels to prevent the mysterious disintegration of their rudder-irons and bolt-heads, though he confessed himself baffled as to the reason the lead caused the corrosion.

The problem recurred when vessels were sheathed in copper to reduce marine weed accumulation and protect against shipworm. In an experiment, the Royal Navy in 1761 had tried fitting the hull of the frigate HMS *Alarm* with 12-ounce copper plating. Upon her return from a voyage to the West Indies, it was found that although the copper remained in fine condition and had indeed deterred shipworm, it had also become detached from the wooden hull in many places because the iron nails used during its installation "...were found dissolved into a kind of rusty Paste". To the surprise of the inspection teams, however, some of the iron nails were virtually undamaged. Closer inspection revealed that water-resistant brown paper trapped under the nail head had inadvertently protected some of the nails: "Where this covering was perfect, the Iron was preserved from Injury". The copper sheathing had been delivered to the dockyard wrapped in the paper which was not always removed before the sheets were nailed to the hull. The conclusion therefore reported to the Admiralty in 1763 was that iron should not be allowed direct contact with copper in sea water.

## US Navy Littoral Combat Ship Independence

Serious galvanic corrosion has been reported on the latest US Navy attack littoral combat vessel the USS *Independence* caused by steel water jet propulsion systems attached to an aluminium hull. Without electrical isolation between the steel and aluminium, the aluminium hull acts as an anode to the stainless steel, resulting in aggressive galvanic corrosion.

## Corroding Lighting Fixtures

The unexpected fall of a heavy light fixture from the ceiling of the Big Dig vehicular tunnel in Boston revealed that corrosion had weakened its support. Improper use of aluminum in contact with stainless steel had caused rapid corrosion in the presence of salt water. The electrochemical potential difference between stainless steel and aluminum is in the range of 0.5 to 1.0V, depending on the exact alloys involved, and can cause considerable corrosion within months under unfavorable conditions. Thousands of failing lights would have to be replaced, at an estimated cost of $54 million.

## Lasagna Cell

A "lasagna cell" is accidentally produced when salty moist food such as lasagna is stored in a steel baking pan and is covered with aluminum foil. After a few hours the foil develops small holes where it touches the lasagna, and the food surface becomes covered with small spots composed of corroded aluminum. In this example, the salty food (lasagna) is the electrolyte, the aluminum foil is the anode, and the steel pan is the cathode. If the aluminum foil only touches the electrolyte in small areas, the galvanic corrosion is concentrated, and corrosion can occur fairly rapidly. If an aluminum baking pan is used instead, the rate of corrosion is markedly reduced, but still may occur.

## Electrolytic Cleaning

The common technique of cleaning silverware by immersion of the silver and a piece of aluminum in an electrolytic bath (usually sodium bicarbonate) is an example of galvanic corrosion. The process involves electrochemical reduction of silver sulfide molecules (generally speaking, silver is not easily corroded by oxygen). Sulfur atoms are stripped off the silver sulfide, transferring them onto and thereby corroding a piece of aluminum foil (a much more reactive metal), leaving pure silver behind. No silver is lost in the process.

## Preventing Galvanic Corrosion

There are several ways of reducing and preventing this form of corrosion.

- Electrically insulate the two metals from each other. If they are not in electrical contact, no galvanic coupling will occur. This can be achieved by using non-conductive materials between metals of different electropotential. Piping can be isolated with a spool of pipe made of plastic materials, or made of metal material internally coated or lined. It is important that the spool be a sufficient length to be effective. For reasons of safety, this should not be attempted where an electrical earthing system uses the pipework for its ground or has equipotential bonding.

- Metal boats connected to a shore line electrical power feed will normally have to have the hull connected to earth for safety reasons. However the end of that earth connection is likely to be a copper rod buried within the marina, resulting in a steel-copper "battery" of about 0.5V. For such cases, the use of a galvanic isolator is essential, typically two semiconductor diodes in series, in parallel with two diodes conducting in the opposite direction. This prevents any current flow while the applied voltage is *less* than 1.4V (i.e. 0.7V per diode), but allows a full flow of current in case of an electrical fault. There will still be a very minor leakage of current through the diodes, which may result in slightly faster corrosion than normal.

- Ensure there is no contact with an electrolyte. This can be done by using water-repellent compounds such as greases, or by coating the metals with an impermeable protective layer, such as a suitable paint, varnish, or plastic. If it is not possible to coat both, the coating should be applied to the more noble, the material with higher potential. This is advisable because if the coating is applied only on the more active material, in case of damage to the coating there will be a large cathode area and a very small anode area, and for the exposed anodic area the corrosion rate will be correspondingly high.

- Using antioxidant paste is beneficial for preventing corrosion between copper and aluminum electrical connections. The paste consists of a lower nobility metal than aluminum or copper.

- Choose metals that have similar electropotentials. The more closely matched the individual potentials, the lesser the potential difference and hence the lesser the galvanic current. Using the same metal for all construction is the easiest way of matching potentials.

- Electroplating or other plating can also help. This tends to use more noble metals that resist corrosion better. Chrome, nickel, silver and gold can all be used. Galvanizing with zinc protects the steel base metal by sacrificial anodic action.

- Cathodic protection uses one or more sacrificial anodes made of a metal which is more active than the protected metal. Alloys of metals commonly used for sacrificial anodes include zinc, magnesium, and aluminium. This approach is commonplace in water heaters and many buried or immersed metallic structures.

- Cathodic Protection can also be applied by connecting a direct current (DC) electrical power supply to oppose the corrosive galvanic current.

Aluminum anodes mounted on a steel-jacketed structure

Electrical panel for a cathodic protection system

## Galvanic Series

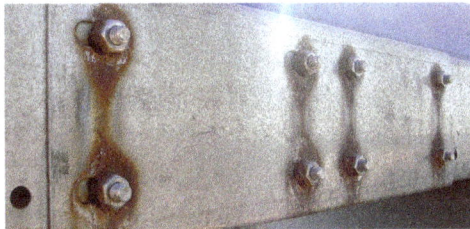

Stainless steel cable ladder with corroding mild steel bolts

All metals can be classified into a galvanic series representing the electrical potential they develop in a given electrolyte against a standard reference electrode. The relative position of two metals on such a series gives a good indication of which metal is more likely to corrode more quickly. However, other factors such as water aeration and flow rate can influence the rate of the process markedly.

## Anodic Index

The compatibility of two different metals may be predicted by consideration of their anodic index. This parameter is a measure of the electrochemical voltage that will be developed between the metal and gold. To find the relative voltage of a pair of metals it is only required to subtract their anodic indices.

This new manifold for water meters has replaced the old one after 4 years of use, having been connected directly to a copper pipe in the building (0.85 V of potential difference, see the point of corrosion in the bottom left). With a PVC battery support, lifetime is unlimited.

Sacrificial anode to protect a boat

For normal environments, such as storage in warehouses or non-temperature and humidity controlled environments, there should not be more than 0.25 V difference in the anodic index. For controlled environments, in which temperature and humidity are controlled, 0.50 V can be tolerated. For harsh environments, such as outdoors, high humidity, and salt environments, there should be not more than 0.15 V difference in the anodic index. For example: gold/silver would have a difference of 0.15V, being acceptable in a harsh environment.

Often when design requires that dissimilar metals come in contact, the galvanic compatibility is managed by finishes and plating. The finishing and plating selected allow the dissimilar materials to be in contact, while protecting the base materials from corrosion. It will always be the metal with the most negative anodic index which will ultimately suffer from corrosion when galvanic incompatibility is in play. This is why sterling silver and stainless steel tableware should never be placed together in a dishwasher at the same time, as the steel items will likely experience corrosion by the end of the cycle (soap and water having served as the chemical electrolyte, and heat having accelerated the process).

| Anodic index | |
|---|---|
| **Metal** | **Index (V)** |
| *Most Cathodic* | |
| Gold, solid and plated, Gold-platinum alloy | −0.00 |
| Rhodium plated on silver-plated copper | −0.05 |
| Silver, solid or plated; monel metal. High nickel-copper alloys | −0.15 |
| Nickel, solid or plated, titanium an s alloys, Monel | −0.30 |
| Copper, solid or plated; low brasses or bronzes; silver solder; German silvery high copper-nickel alloys; nickel-chromium alloys | −0.35 |
| Brass and bronzes | −0.40 |
| High brasses and bronzes | −0.45 |
| 18% chromium type corrosion-resistant steels | −0.50 |
| Chromium plated; tin plated; 12% chromium type corrosion-resistant steels | −0.60 |
| Tin-plate; tin-lead solder | −0.65 |
| Lead, solid or plated; high lead alloys | −0.70 |
| 2000 series wrought aluminum | −0.75 |
| Iron, wrought, gray or malleable, plain carbon and low alloy steels | −0.85 |
| Aluminum, wrought alloys other than 2000 series aluminum, cast alloys of the silicon type | −0.90 |
| Aluminum, cast alloys other than silicon type, cadmium, plated and chromate | −0.95 |
| Hot-dip-zinc plate; galvanized steel | −1.20 |
| Zinc, wrought; zinc-base die-casting alloys; zinc plated | −1.25 |
| Magnesium & magnesium-base alloys, cast or wrought | −1.75 |
| Beryllium | −1.85 |
| *Most Anodic* | |

# Crevice Corrosion

Crevice corrosion refers to corrosion occurring in confined spaces to which the access of the working fluid from the environment is limited. These spaces are generally called crevices. Examples of crevices are gaps and contact areas between parts, under gaskets or seals, inside cracks and seams, spaces filled with deposits and under sludge piles.

Corrosion can be occurred anywhere like, in the crevice between the tube and tube sheet (both made of type 316 stainless steel) of a heat exchanger in a sea water desalination plant.

## Mechanism

The corrosion resistance of a stainless steel is dependent on the presence of an ultra-thin protective oxide film (passive film) on its surface, but it is possible under certain conditions for this oxide film to break down, for example in halide solutions or reducing acids. Areas where the oxide film can break down can also sometimes be the result of the way components are designed, for example under gaskets, in sharp re-entrant corners or associated with incomplete weld penetration or overlapping surfaces. These can all form crevices which can promote corrosion. To function as a corrosion site, a crevice has to be of sufficient width to permit entry of the corrodent, but narrow enough to ensure that the corrodent remains stagnant. Accordingly crevice corrosion usually occurs in gaps a few micrometres wide, and is not found in grooves or slots in which circulation of the corrodent is possible. This problem can often be overcome by paying attention to the design of the component, in particular to avoiding formation of crevices or at least keeping them as open as possible. Crevice corrosion is a very similar mechanism to pitting corrosion; alloys resistant to one are generally resistant to both. Crevice corrosion can be viewed as a less severe form of localized corrosion when compared with pitting. The depth of penetration and the rate of propagation in pitting corrosion are significantly greater than in crevice corrosion.

Crevices can develop a local chemistry which is very different from that of the bulk fluid. For example, in boilers, concentration of non-volatile impurities may occur in crevices near heat-transfer surfaces because of the continuous water vaporization. "Concentration factors" of many millions are not uncommon for common water impurities like sodium, sulfate or chloride. The concentration process is often referred to as "hideout" (HO), whereas the opposite process, whereby the concentrations tend to even out (e.g., during shutdown) is called "hideout return" (HOR). In a neutral pH solution, the pH inside the crevice can drop to 2, a highly acidic condition that accelerates the corrosion of most metals and alloys.

For a given crevice type, two factors are important in the initiation of crevice corrosion: the chemical composition of the electrolyte in the crevice and the potential drop into the crevice. Researchers had previously claimed that either one or the other of the two factors was responsible for initiating crevice corrosion, but recently it has been shown that it is a combination of the two that causes active crevice corrosion. Both the potential drop and the change in composition of the crevice electrolyte are caused by deoxygenation of the crevice and a separation of electroactive areas, with net anodic reactions occurring within the crevice and net cathodic reactions occurring exterior to the crevice (on the bold surface). The ratio of the surface areas between the cathodic and anodic region is significant.

Some of the phenomena occurring within the crevice may be somewhat reminiscent of galvanic corrosion:

galvanic corrosion

    two connected metals + single environment

crevice corrosion

> one metal part + two connected environments

The mechanism of crevice corrosion can be (but is not always) similar to that of pitting corrosion. However, there are sufficient differences to warrant a separate treatment. For example, in crevice corrosion, one has to consider the geometry of the crevice and the nature of the concentration process leading to the development of the differential local chemistry. The extreme and often unexpected local chemistry conditions inside the crevice need to be considered. Galvanic effects can play a role in crevice degradation.

## Mode of Attack

Depending on the environment developed in the crevice and the nature of the metal, the crevice corrosion can take a form of:

- pitting (i.e., formation of pits),

- filiform corrosion (this type of crevice corrosion that may occur on a metallic surface underneath an organic coating),

- intergrannular attack, or

- stress corrosion cracking.

## Stress Corrosion Cracking

The Silver Bridge upon completion in 1928

A common form of crevice failure occurs due to stress corrosion cracking, where a crack or cracks develop from the base of the crevice where the stress concentration is greatest. This was the root cause of the fall of the Silver Bridge in 1967 in West Virginia, where a single critical crack only about 3 mm long suddenly grew and fractured a tie bar joint. The rest of the bridge fell in less than a minute. The eyebars in the Silver Bridge were not redundant, as links were composed of only two bars each, of high strength steel (more than twice as strong as common mild steel), rather than a thick stack of thinner bars of modest material strength "combed" together as is usual for redundancy. With only two bars, the failure of one could impose excessive loading on the second, causing total failure—unlikely if more bars are used. While a low-redundancy chain can be engineered to the design requirements, the safety is completely dependent upon correct, high quality manufacturing and assembly.

## Significance

The susceptibility to crevice corrosion varies widely from one material-environment system to another. In general, crevice corrosion is of greatest concern for materials which are normally passive metals, like stainless steel or aluminum. Crevice corrosion tends to be of greatest significance to components built of highly corrosion-resistant superalloys and operating with the purest-available water chemistry. For example, steam generators in nuclear power plants degrade largely by crevice corrosion.

Crevice corrosion is extremely dangerous because it is localized and can lead to component failure while the overall material loss is minimal. The initiation and progress of crevice corrosion can be difficult to detect.

## Microbial Corrosion

Microbial corrosion, also called bacterial corrosion, bio-corrosion, microbiologically influenced corrosion, or microbially induced corrosion (MIC), is corrosion caused or promoted by microorganisms, usually chemoautotrophs. It can apply to both metals and non-metallic materials.

## Bacteria

Some sulfate-reducing bacteria produce hydrogen sulfide, which can cause sulfide stress cracking. *Acidithiobacillus* bacteria produce sulfuric acid; *Acidothiobacillus thiooxidans* frequently damages sewer pipes. *Ferrobacillus ferrooxidans* directly oxidizes iron to iron oxides and iron hydroxides; the rusticles forming on the RMS *Titanic* wreck are caused by bacterial activity. Other bacteria produce various acids, both organic and mineral, or ammonia.

In presence of oxygen, aerobic bacteria like *Acidithiobacillus thiooxidans*, *Thiobacillus thioparus*, and *Thiobacillus concretivorus*, all three widely present in the environment, are the common corrosion-causing factors resulting in biogenic sulfide corrosion.

Without presence of oxygen, anaerobic bacteria, especially *Desulfovibrio* and *Desulfotomaculum*, are common. *Desulfovibrio salixigens* requires at least 2.5% concentration of sodium chloride, but *D. vulgaris* and *D. desulfuricans* can grow in both fresh and salt water. *D. africanus* is another common corrosion-causing microorganism. The *Desulfotomaculum* genus comprises sulfate-reducing spore-forming bacteria; *Dtm. orientis* and *Dtm. nigrificans* are involved in corrosion processes. Sulfate-reducers require reducing environment; an electrode potential lower than -100 mV is required for them to thrive. However, even a small amount of produced hydrogen sulfide can achieve this shift, so the growth, once started, tends to accelerate.

Layers of anaerobic bacteria can exist in the inner parts of the corrosion deposits, while the outer parts are inhabited by aerobic bacteria.

Some bacteria are able to utilize hydrogen formed during cathodic corrosion processes.

Bacterial colonies and deposits can form concentration cells, causing and enhancing galvanic corrosion.

Bacterial corrosion may appear in form of pitting corrosion, for example in pipelines of the oil and gas industry. Anaerobic corrosion is evident as layers of metal sulfides and hydrogen sulfide smell. On cast iron, a graphitic corrosion selective leaching may be the result, with iron being consumed by the bacteria, leaving graphite matrix with low mechanical strength in place.

Various corrosion inhibitors can be used to combat microbial corrosion. Formulae based on benzalkonium chloride are common in oilfield industry.

Microbial corrosion can also apply to plastics, concrete, and many other materials. Two examples are Nylon-eating bacteria and Plastic-eating bacteria.

## Aviation Fuel

Hydrocarbon utilizing microorganisms, mostly *Cladosporium resinae* and *Pseudomonas aeruginosa* and Sulfate Reducing Bacteria, colloquially known as "HUM bugs", are commonly present in jet fuel. They live in the water-fuel interface of the water droplets, form dark black/brown/green, gel-like mats, and cause microbial corrosion to plastic and rubber parts of the aircraft fuel system by consuming them, and to the metal parts by the means of their acidic metabolic products. They are also incorrectly called algae due to their appearance. FSII, which is added to the fuel, acts as a growth retardant for them. There are about 250 kinds of bacteria that can live in jet fuel, but fewer than a dozen are meaningfully harmful.

## Nuclear Waste

Microorganisms can affect negatively radio elements confinement in nuclear waste.

## Sewerage

Sewer network structures are prone to biodeterioration of materials due to the action of some microorganisms associated to the sulfur cycle. It can be a severely damaging phenomenon which was firstly described by Olmstead and Hamlin in 1900 for a brick sewer located in Los Angeles. Jointed mortar between the bricks disintegrated and ironwork was heavily rusted. The mortar joint had ballooned to two to three times its original volume, leading to the destruction or the loosening of some bricks.

Around 9% of damages described in sewer networks can be ascribed to the successive action of two kinds of microorganisms. Sulfate-reducing bacteria (SRB) can grow in relatively thick layers of sedimentary sludge and sand (typically 1 mm thick) accumulating at the bottom of the pipes and characterized by anoxic conditions. They can grow using oxidized sulfur compounds present in the effluent as electron acceptor and excrete hydrogen sulfide ($H_2S$). This gas is then emitted in the aerial part of the pipe and can impact the structure in two ways: either directly by reacting with the material and leading to a decrease in pH, or indirectly through its use as a nutrient by sulfur-oxidizing bacteria (SOB), growing in oxic conditions, which produce biogenic sulfuric acid. The structure is then submitted to a biogenic sulfuric acid attack. Materials like calcium aluminate cements, PVC or vitrified clay pipe may be substituted for ordinary concrete or steel sewers that are not resistant in these environments.

## Anaerobic Corrosion

Hydrogen corrosion is a form of metal corrosion occurring in the presence of anoxic water. Hydrogen corrosion involves a redox reaction that reduces hydrogen ions, forming molecular hydrogen.

Metals enter aqueous solution and are oxidized.

*Oxidation reaction (pH independent):*

$$Fe \rightarrow Fe^{2+} + 2e^-$$

*Reduction reaction in acid solution:*

$$2H^+ + 2e^- \rightarrow H_2$$

In an acidic solution, the water molecules are protonated and the hydronium ions ($H_3O^+$) are directly reduced into $H_2$.

*Reduction reaction in neutral or slightly alkaline solution:*

$$2H_2O + 2e^- \rightarrow H_2 + 2OH^-$$

In a neutral or slightly alkaline solution, the protons of water are reduced into molecular hydrogen giving rise to the production of hydroxide ions responsible of the precipitation of the slightly soluble ferrous hydroxide ($Fe(OH)_2$).

This finally leads to the global reaction of the anaerobic corrosion of iron in water:

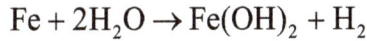

$$Fe + 2H_2O \rightarrow Fe(OH)_2 + H_2$$

## Transformation of Ferrous Hydroxide Into Magnetite

Under anaerobic conditions, the ferrous hydroxide ($Fe(OH)_2$) can be oxidized by the protons of water to form magnetite and molecular hydrogen. This process is described by the Schikorr reaction:

$$3\ Fe(OH)_2 \rightarrow Fe_3O_4 + H_2 + 2\ H_2O$$

*ferrous hydroxide $\rightarrow$ magnetite + hydrogen + water*

The well crystallized magnetite ($Fe_3O_4$) is thermodynamically more stable than the ferrous hydroxide ($Fe(OH)_2$).

This process also occurs during the anaerobic corrosion of iron and steel in oxygen-free groundwater and in reducing soils below the water table.

# Corrosion in Ballast Tanks

Corrosion in Ballast Tanks is the deterioration process where the surface of a ballast tank progresses from microblistering, to electroendosmotic blistering, and finally to cracking of the tank steel itself.

> "Effective corrosion control in segregated water ballast spaces is probably the single most important feature, next to the integrity of the initial design, in determining the ship's effective life span and structural reliability," as said by Germanischer Lloyd's Principal surveyor.

Throughout the years the merchant fleet has become increasingly aware of the importance of avoiding corrosion in ballast tanks.

## Factors with an Influence on Corrosion in Ballast Tanks

Epoxy and modified epoxy are standard coatings used to provide protective barriers to corrosion in ballast tanks. Exposed, unprotected steel will corrode much more rapidly than steel covered with this protective layer. Many ships also use sacrificial anodes or an impressed current for additional protection. Empty ballast tanks will corrode faster than areas fully immersed due to the thin - and electo conducting - moisture film covering them.

The main factors influencing the rate of corrosion are diffusion, temperature, Conductivity, type of ions, pH, and electrochemical corrosion potential.

## Regions of a Ballast Tank

Ballast tanks do not corrode uniformly throughout the tank. Each region behaves distinctively, according to it electrochemical loading. The differences can especially be seen in empty ballast tanks. The upper sections usually corrode but the lower sections will blister.

A ballast tank has three distinct sections: 1) upper, 2) mid or "boottop" area and, 3) the "double bottom" or lower wing sections. The upper regions are constantly affected by weather. This area experiences a high degree of thermal cycling and mechanical damage through vibration. This area tends to undergo anodic oxidation more rapidly than other sections and will weaken more rapidly. This ullage or headspace area contains more oxygen and thus speeds atmospheric corrosion, as evidenced by the appearance of rust scales.

In the midsection corrodes more slowly than upper or the bottom sections of the tank.

Double bottoms are prone to cathodic blistering. Temperatures in this area are much lower due to the cooling of the sea. If this extremely cathodic region is placed close to an anodic source (e.g. a corroding ballast pipe), cathodic blistering may occur especially where the epoxy coating is relatively new. Mud retained in ballast water can lead to microial corrosion.

## Marine Regulations

Many maritime accidents have been caused by corrosion, and this has led to stringent regulations concerning protective coatings for ballast tanks. *The Coating Performance Standard for Ballast Tank Coatings* (PSPC), became effective in 2008. It specifies how protective coatings should be applied during vessel construction with the intention of giving a coating a 15-year service life. Additional regulations, such as those established by *The International Convention for the Control and Management of Ships Ballast Water & Sediments* (SBWS) sought to avoid introducing invasive species throughout the world through ship's ballast tanks. The methods used to avoid having these invasive species surviving in ballast tanks however greatly increased the rate of corrosion. Therefore ongoing research attempts to find water treatment systems that kill invasive species, while not having a destructive effect on the ballast tank coatings. As double-hulled tankers were introduced it meant that there was more ballast tank area had to be coated and therefore a greater capital investment for ship owners. With the onset of the OPA 90 and later the amendments to MARPOL annex 1, single hull tankers (without alternative method) have basically phased out.

Modern double hull tankers, with their fully "segregated ballast tanks" propose another problem. Empty tanks act as insulation from the cold sea and allow the warm cargo areas to retain their heat longer. Corrosion rates increase with differences in temperature. Consequently, the cargo side of the ballast tank corrodes more quickly than it did with single hull tankers.

## Corrosion in Space

Corrosion in space is the corrosion of materials occurring in outer space. Instead of moisture and oxygen acting as the primary corrosion causes, the materials exposed to outer space are subjected to vacuum, bombardment by ultraviolet and X-rays, and high-energy charged particles (mostly electrons and protons from solar wind). In the upper layers of the atmosphere (between 90–800 km), the atmospheric atoms, ions, and free radicals, most notably atomic oxygen, play a major role. The concentration of atomic oxygen depends on altitude and solar activity, as the bursts of ultraviolet radiation cause photodissociation of molecular oxygen. Between 160 and 560 km, the atmosphere consists of about 90% atomic oxygen.

### Materials

Corrosion in space has the highest impact on spacecraft with moving parts. Early satellites tended to develop problems with seizing bearings. Now the bearings are coated with a thin layer of gold.

Different materials resist corrosion in space differently. For example, aluminium is slowly eroded by atomic oxygen, while gold and platinum are highly corrosion-resistant. Gold-coated foils and thin layers of gold on exposed surfaces are therefore used to protect the spacecraft from the harsh environment. Thin layers of silicon dioxide deposited on the surfaces can also protect metals from the effects of atomic oxygen; e.g., the Starshine 3 satellite aluminium front mirrors were protected that way. However, the protective layers are subject to erosion by micrometeorites.

Silver builds up a layer of silver oxide, which tends to flake off and has no protective function; such gradual erosion of silver interconnects of solar cells was found to be the cause of some observed in-orbit failures.

Many plastics are considerably sensitive to atomic oxygen and ionizing radiation. Coatings resistant to atomic oxygen are a common protection method, especially for plastics. Silicone-based paints and coatings are frequently employed, due to their excellent resistance to radiation and atomic oxygen. However, the silicone durability is somewhat limited, as the surface exposed to atomic oxygen is converted to silica which is brittle and tends to crack.

## Solving Corrosion

The process of space corrosion is being actively investigated. One of the efforts aims to design a sensor based on zinc oxide, able to measure the amount of atomic oxygen in the vicinity of the spacecraft; the sensor relies on drop of electrical conductivity of zinc oxide as it absorbs further oxygen.

## Other Problems

The outgassing of volatile silicones on low Earth orbit devices leads to presence of a cloud of contaminants around the spacecraft. Together with atomic oxygen bombardment, this may lead to gradual deposition of thin layers of carbon-containing silicon dioxide. Their poor transparency is a concern in case of optical systems and solar panels. Deposits of up to several micrometers were observed after 10 years of service on the solar panels of the Mir space station.

Other sources of problems for structures subjected to outer space are erosion and re-deposition of the materials by sputtering caused by fast atoms and micrometeorites. Another major concern, though of non-corrosive kind, is material fatigue caused by cyclical heating and cooling and associated thermal expansion mechanical stresses.

# Biogenic Sulfide Corrosion

Biogenic sulfide corrosion is a bacterially mediated process of forming hydrogen sulfide gas and the subsequent conversion to sulfuric acid that attacks concrete and steel within wastewater environments. The hydrogen sulfide gas is biochemically oxidized in the presence of moisture to form sulfuric acid. The effect of sulfuric acid on concrete and steel surfaces exposed to severe wastewater environments can be devastating. In the USA alone, corrosion is causing sewer asset losses estimated at around $14 billion per year. This cost is expected to increase as the aging infrastructure continues to fail.

## Environment

Corrosion may occur where stale sewage generates hydrogen sulfide gas into an atmosphere containing oxygen gas and high relative humidity. There must be an underlying anaerobic aquatic habitat containing sulfates and an overlying aerobic aquatic habitat separated by a gas phase containing both oxygen and hydrogen sulfide at concentrations in excess of 2 ppm.

## Conversion of Sulfate $SO_4^{2-}$ to Hydrogen Sulfide $H_2S$

Fresh domestic sewage entering a wastewater collection system contains proteins including organic sulfur compounds oxidizable to sulfates and may contain inorganic sulfates. Dissolved oxygen is depleted as bacteria begin to catabolize organic material in sewage. In the absence of dissolved oxygen and nitrates, sulfates are reduced to hydrogen sulfide as an alternative source of oxygen for catabolizing organic waste by sulfate reducing bacteria (SRB), identified primarily from the obligate anaerobic species *Desulfovibrio*.

Hydrogen sulfide production depends on various physicochemical, topographic and hydraulic parameters such as:

- Sewage oxygen concentration. The threshold is 0.1 mg.l$^{-1}$; above this value, sulfides produced in sludge and sediments are oxidized by oxygen; below this value, sulfides are emitted in the gaseous phase.

- Temperature. The higher the temperature, the faster the kinetics of $H_2S$ production.

- Sewage pH. It must be included between 5.5 and 9 with an optimum at 7.5-8.

- Sulfate concentration.

- Nutrients concentration, associated to the biochemical oxygen demand.

- Conception of the sewerage As $H_2S$ is formed only in anaerobic conditions, slow flow and long retention time gives more time to aerobic bacteria to consume all available dissolved oxygen in water, creating anaerobic conditions. The flatter the land, the less slope can be given to the sewer network, and this favors slower flow and more pumping stations (where retention time is generally longer)

## Conversion of Hydrogen Sulfide to Sulfuric acid $H_2SO_4$

Some hydrogen sulfide gas diffuses into the headspace environment above the wastewater. Moisture evaporated from warm sewage may condense on unsubmerged walls of sewers, and is likely to hang in partially formed droplets from the horizontal crown of the sewer. As a portion of the hydrogen sulfide gas and oxygen gas from the air above the sewage dissolves into these stationary droplets, they become a habitat for sulfur oxidizing bacteria (SOB), of the genus *Acidithiobacillus*. Colonies of these aerobic bacteria metabolize the hydrogen sulfide gas to sulfuric acid.

## Corrosion

Sulfuric acid produced by microorganisms will interact with the surface of the structure material. For ordinary Portland cement, it reacts with the calcium hydroxide in concrete to form calcium sulfate. This change simultaneously destroys the polymeric nature of calcium hydroxide and substitutes a larger molecule into the matrix causing pressure and spalling of the adjacent concrete and aggregate particles. The weakened crown may then collapse under heavy overburden loads. Even within a well-designed sewer network, a rule of thumb in the industry suggests that 5% of the total length may/will suffer from biogenic corrosion. In these specific areas, biogenic sulfide corrosion can deteriorate metal or several millimeters per year of concrete.

| Source | Thickness loss (in mm.y$^{-1}$) | Material type |
|---|---|---|
| US EPA, 1991 | 2.5 − 10 | Concrete |
| Morton et al., 1991 | 2.7 | Concrete |
| Mori et al., 1992 | 4.3 − 4.7 | Concrete |
| Ismail et al., 1993 | 2 − 4 | Mortar |
| Davis, 1998 | 3.1 | Concrete |
| Monteny et al., 2001 | 1.0 − 1.3 | Mortar |
| Vincke et al., 2002 | 1.1 − 1.8 | Concrete |

For calcium aluminate cements, processes are completely different because they are based on another chemical composition. At least three different mechanisms contribute to the better resistance to biogenic corrosion:

- The first barrier is the larger acid neutralizing capacity of calcium aluminate cements vs. ordinary Portland Cement; one gram of calcium aluminate cement can neutralize around 40% more acid than a gram of ordinary Portland Cement. For a given production of acid by the biofilm, a calcium aluminate cement concrete will last longer.

- The second barrier is due to the precipitation, when the surficial pH gets below 10, of a layer of alumina gel (AH3 in cement chemistry notation). AH3 is a stable compound down to a pH of 4 and it will form an acid-resistant barrier as long as the surface pH is not lowered below 3-4 by the bacterial activity.

- The third barrier is the bacteriostatic effect locally activated when the surface reaches pH values less than 3-4. At this level, the alumina gel is no longer stable and will dissolve, liberating aluminum ions. These ions will accumulate in the thin biofilm. Once the concentration reaches 300-500 ppm, it will produce a bacteriostatic effect on bacteria metabolism. In other word, bacteria will stop oxidizing the sulfur from $H_2S$ to produce acid, and the pH will stop decreasing.

A mortar made of calcium aluminate cement combined with calcium aluminate aggregates, i.e. a 100% calcium aluminate material, will last much longer as aggregates can also limit microorganisms' growth and inhibits the acid generation at the source itself.

## Prevention

There are several options to address biogenic sulfide corrosion problems: impairing $H_2S$ formation, venting out the $H_2S$ or using materials resistant to biogenic corrosion. For example, sewage flows more rapidly through steeper gradient sewers reducing time available for hydrogen sulfide generation. Likewise, removing sludge and sediments from the bottom of the pipes reduces the amount of anoxic areas responsible for sulfate reducing bacteria growth. Providing good ventilation of sewers can reduce atmospheric concentrations of hydrogen sulfide gas and may dry exposed sewer crowns, but this may create odor issues with neighbors around the venting shafts. Three other efficient methods can be used involving continuous operation of mechanical equipment: chemical reactant like calcium nitrate can be continuously added in the sewerage water to impair the $H_2S$ formation, an active ventilation through odor treatment units to remove $H_2S$, or an injection of compressed air in pressurized mains to avoid the anaerobic condition to develop. In sewerage areas where biogenic sulfide corrosion is expected, acid resistant materials like calcium aluminate cements, PVC or vitrified clay pipe may be substituted to ordinary concrete or steel sewers. Existing structures that have extensive exposure to biogenic corrosion such as sewer manholes and pump station wet wells can be rehabilitated. Rehabilitation can be done with materials such as a structural epoxy coating, this epoxy is designed to be both acid resistant and strength the compromised concrete structure.

## High-temperature Corrosion

High-temperature sulfur corrosion of a 12 CrMo 19 5 pipe stub

High-temperature corrosion is a mechanism of corrosion that takes place in gas turbines, diesel engines, furnaces or other machinery coming in contact with hot gas containing certain contaminants. Fuel sometimes contains vanadium compounds or sulfates which can form compounds during combustion having a low melting point. These liquid melted salts are strongly corrosive for stainless steel and other alloys normally inert against the corrosion and high temperatures. Other high-temperature corrosions include high-temperature oxidation, sulfidation and carbonization.

### Sulfates

Two types of sulfate-induced hot corrosion are generally distinguished: Type I takes place above the melting point of sodium sulfate and Type II occurs below the melting point of sodium sulfate but in the presence of small amounts of $SO_3$.

In Type I the protective oxide scale is dissolved by the molten salt. Sulfur is released from the salt and diffuses into the metal substrate forming discrete grey/blue colored aluminum or chromium sulfides so that, after the salt layer has been removed, the steel cannot rebuild a new protective oxide layer. Alkali sulfates are formed from sulfur trioxide and sodium-containing compounds. As the formation of vanadates is preferred, sulfates are formed only if the amount of alkali metals is higher than the corresponding amount of vanadium.

The same kind of attack has been observed for potassium and magnesium sulfate.

## Vanadium

Vanadium is present in petroleum, especially from Canada, western United States, Venezuela and the Caribbean region, in the form of porphyrine complexes. These complexes get concentrated on the higher-boiling fractions, which are the base of heavy residual fuel oils. Residues of sodium, primarily from sodium chloride and spent oil treatment chemicals, are also present. More than 100 ppm of sodium and vanadium will yield ash capable of causing fuel ash corrosion.

Most fuels contain small traces of vanadium. The vanadium is oxidized to different vanadates. Molten vanadates present as deposits on metal can flux oxide scales and passivation layers. Furthermore, the presence of vanadium accelerates the diffusion of oxygen through the fused salt layer to the metal substrate; vanadates can be present in semiconducting or ionic form, where the semiconducting form has significantly higher corrosivity as the oxygen is transported via oxygen vacancies. Ionic form in contrast transports oxygen by diffusion of the vanadates, which is significantly slower. The semiconducting form is rich on vanadium pentoxide.

At high temperatures or lower availability of oxygen, refractory oxides - vanadium dioxide and vanadium trioxide - form. These do not promote corrosion. However, at conditions most common for burning, vanadium pentoxide gets formed. Together with sodium oxide, vanadates of various composition ratios are formed. Vanadates of composition approximating $Na_2O.6\,V_2O_5$ have the highest corrosion rates at the temperatures between 593 °C and 816 °C; at lower temperatures the vanadate is in solid state, at higher temperatures vanadates with higher proportion of vanadium provide higher corrosion rates.

The solubility of the passivation layer oxides in the molten vanadates depends on the composition of the oxide layer. Iron(III) oxide is readily soluble in vanadates between $Na_2O.6\,V_2O_5$ and $6\,Na_2O.V_2O_5$, at temperatures below 705 °C in amounts up to equal to the mass of the vanadate. This composition range is common for ashes, which aggravates the problem. Chromium(III) oxide, nickel(II) oxide, and cobalt(II) oxide are less soluble in vanadates; they convert the vanadates to less corrosive ionic form and their vanadates are tightly adherent, refractory, and acting as oxygen barriers.

The corrosion rate by vanadates can be lowered by lowering the amount of excess air for combustion (thus forming preferentially the refractory oxides), refractory coatings of the exposed surfaces, or use of high-chromium alloys, e.g. 50% Ni/50% Cr or 40% Ni/60% Cr.

The presence of sodium in a ratio of 1:3 gives the lowest melting point and must be avoided. This melting point of 535 °C can cause problems on the hot spots of the engine like piston crowns, valve seats, and turbochargers.

## Lead

Lead can form a low melting slag capable of fluxing protective oxide scales.

## Intergranular Corrosion

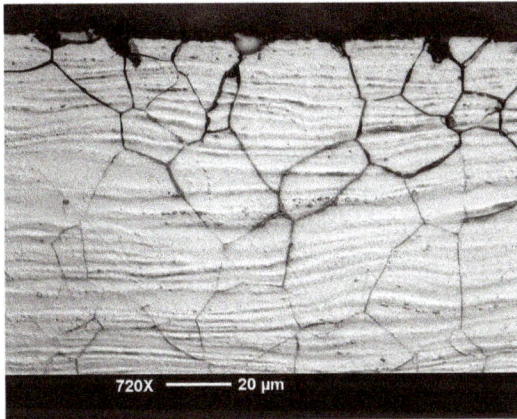

Microscope view of a polished cross section of a
material attacked by intergranular corrosion

Intergranular corrosion (IGC), also known as intergranular attack (IGA), is a form of corrosion where the boundaries of crystallites of the material are more susceptible to corrosion than their insides. (*Cf.* transgranular corrosion.)

This situation can happen in otherwise corrosion-resistant alloys, when the grain boundaries are depleted, known as *grain boundary depletion*, of the corrosion-inhibiting elements such as chromium by some mechanism. In nickel alloys and austenitic stainless steels, where chromium is added for corrosion resistance, the mechanism involved is precipitation of chromium carbide at the grain boundaries, resulting in the formation of chromium-depleted zones adjacent to the grain boundaries (this process is called sensitization). Around 12% chromium is minimally required to ensure passivation, a mechanism by which an ultra thin invisible film, known as passive film, forms on the surface of stainless steels. This passive film protects the metal from corrosive environments. The self-healing property of the passive film make the steel stainless. Selective leaching often involves grain boundary depletion mechanisms.

These zones also act as local galvanic couples, causing local galvanic corrosion. This condition happens when the material is heated to temperature around 700 °C for too long time, and often happens during welding or an improper heat treatment. When zones of such material form due to welding, the resulting corrosion is termed weld decay. Stainless steels can be stabilized against this behavior by addition of titanium, niobium, or tantalum, which form titanium carbide, niobium carbide and tantalum carbide preferentially to chromium carbide, by lowering the content of carbon in the steel and in case of welding also in the filler metal under 0.02%, or by heating the entire part above 1000 °C and quenching it in water, leading to dissolution of the chromium carbide in the grains and then preventing its precipitation. Another possibility is to keep the welded parts thin enough so that, upon cooling, the metal dissipates heat too quickly for chromium carbide to precipitate. The ASTM A923, ASTM A262, and other similar tests are often used to determine when stainless steels are susceptible to intergranular corrosion. The tests require etching with chemicals that reveal the presence of intermetallic particles, sometimes combined with Charpy V-Notch and other mechanical testing.

Another related kind of intergranular corrosion is termed knifeline attack (KLA). Knifeline attack impacts steels stabilized by niobium, such as 347 stainless steel. Titanium, niobium, and their carbides dissolve in steel at very high temperatures. At some cooling regimes (depending on the rate of cooling), niobium carbide does not precipitate and the steel then behaves like unstabilized steel, forming chromium carbide instead. This affects only a thin zone several millimeters wide in the very vicinity of the weld, making it difficult to spot and increasing the corrosion speed. Structures made of such steels have to be heated in a whole to about 1065 °C (1950 °F), when the chromium carbide dissolves and niobium carbide forms. The cooling rate after this treatment is not important, as the carbon that would otherwise pose risk of formation of chromium carbide is already sequestered as niobium carbide.

Aluminium-based alloys may be sensitive to intergranular corrosion if there are layers of materials acting as anodes between the aluminium-rich crystals. High strength aluminium alloys, especially when extruded or otherwise subjected to high degree of working, can undergo exfoliation corrosion, where the corrosion products build up between the flat, elongated grains and separate them, resulting in lifting or leafing effect and often propagating from edges of the material through its entire structure. Intergranular corrosion is a concern especially for alloys with high content of copper.

Other kinds of alloys can undergo exfoliation as well; the sensitivity of cupronickel increases together with its nickel content. A broader term for this class of corrosion is lamellar corrosion. Alloys of iron are susceptible to lamellar corrosion, as the volume of iron oxides is about seven times higher than the volume of original metal, leading to formation of internal tensile stresses tearing the material apart. Similar effect leads to formation of lamellae in stainless steels, due to the difference of thermal expansion of the oxides and the metal.

Copper-based alloys become sensitive when depletion of copper content in the grain boundaries occurs.

Anisotropic alloys, where extrusion or heavy working leads to formation of long, flat grains, are especially prone to intergranular corrosion.

Intergranular corrosion induced by environmental stresses is termed stress corrosion cracking. Inter granular corrosion can be detected by ultrasonic and eddy current methods.

## Sensitization Effect

*Sensitization* refers to the precipitation of carbides at grain boundaries in a stainless steel or alloy, causing the steel or alloy to be susceptible to intergranular corrosion or intergranular stress corrosion cracking.

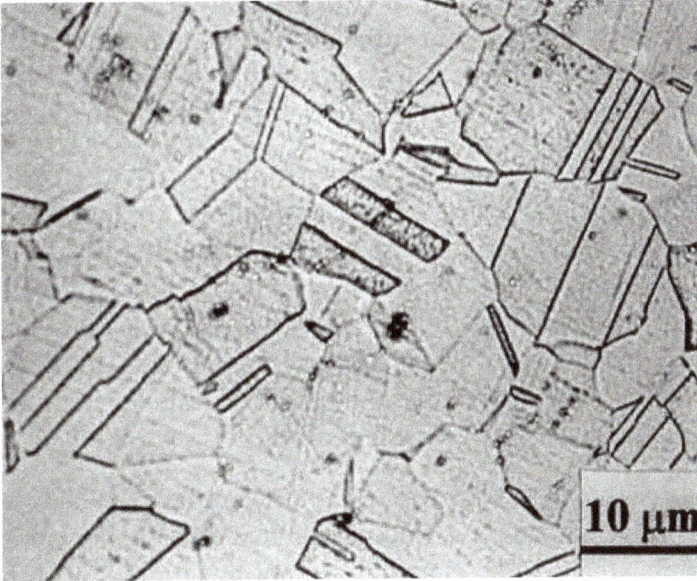

Unsensitized microstructure

Certain alloys when exposed to a temperature characterized as a sensitizing temperature become particularly susceptible to intergranular corrosion. In a corrosive atmosphere, the grain interfaces of these sensitized alloys become very reactive and intergranular corrosion results. This is characterized by a localized attack at and adjacent to grain boundaries with relatively little corrosion of the grains themselves. The alloy disintegrates (grains fall out) and/or loses its strength.

The photos show the typical microstructure of a normalized (unsensitized) type 304 stainless steel and a heavily sensitized steel. The samples have been polished and etched before taking the photos, and the sensitized areas show as wide, dark lines where the etching fluid has caused corrosion. The dark lines consist of carbides and corrosion products.

Intergranular corrosion is generally considered to be caused by the segregation of impurities at the grain boundaries or by enrichment or depletion of one of the alloying elements in the grain boundary areas. Thus in certain aluminium alloys, small amounts of iron have been shown to segregate in the grain boundaries and cause intergranular corrosion. Also, it has been shown that the zinc content of a brass is higher at the grain boundaries and subject to such corrosion. High-strength aluminium alloys such as the Duralumin-type alloys (Al-Cu) which depend upon precipitated phases for strengthening are susceptible to intergranular corrosion following sensitization at temperatures of about 120 °C. Nickel-rich alloys such as Inconel 600 and Incoloy 800 show similar susceptibility. Die-cast zinc alloys containing aluminum exhibit intergranular corrosion by steam in a marine atmosphere. Cr-Mn and Cr-Mn-Ni steels are also susceptible to intergranular corrosion following sensitization in the temperature range of 420°-850 °C. In the case of the austenitic stainless steels, when these steels are sensitized by being heated in the temperature range of about 520° to 800 °C, depletion of chromium in the grain boundary region occurs, resulting in susceptibility to intergranular corrosion. Such sensitization of austenitic stainless steels can readily occur because of temperature service requirements, as in steam generators, or as a result of subsequent welding of the formed structure.

Several methods have been used to control or minimize the intergranular corrosion of susceptible alloys, particularly of the austenitic stainless steels. For example, a high-temperature solution heat treatment, commonly termed solution-annealing, quench-annealing or solution-quenching, has been used. The alloy is heated to a temperature of about 1,060° to 1,120 °C and then water quenched. This method is generally unsuitable for treating large assemblies, and also ineffective where welding is subsequently used for making repairs or for attaching other structures.

Another control technique for preventing intergranular corrosion involves incorporating strong carbide formers or stabilizing elements such as niobium or titanium in the stainless steels. Such elements have a much greater affinity for carbon than does chromium; carbide formation with these elements reduces the carbon available in the alloy for formation of chromium carbides. Such a stabilized titanium-bearing austenitic chromium-nickel-copper stainless steel is shown in U.S. Pat. No. 3,562,781. Or the stainless steel may initially be reduced in carbon content below 0.03 percent so that insufficient carbon is provided for carbide formation. These techniques are expensive and only partially effective since sensitization may occur with time. The low-carbon steels also frequently exhibit lower strengths at high temperatures.

# Metal Dusting

Metal dusting is "a catastrophic form of corrosion that occurs when susceptible materials are exposed to environments with high carbon activities." The corrosion manifests itself as a break-up of bulk metal to metal powder. The suspected mechanism is firstly the deposition of a graphite layer on the surface of the metal, usually from carbon monoxide (CO) in the vapour phase. This graphite layer is then thought to form metastable $M_3C$ species (where M is the metal), which migrate away from the metal surface. However, in some regimes no $M_3C$ species are observed indicating a direct transfer of metal atoms into the graphite layer.

The temperatures normally associated with metal dusting are high (300–850 °C). From a general understanding of chemistry, it can be deduced that at lower temperatures, the rate of reaction to form the metastable $M_3C$ species is too low to be significant, and at much higher temperatures the graphite layer is unstable and so CO deposition does not occur (at least to any appreciable degree).

Very briefly, there are several proposed methods for prevention or reduction of metal dusting; the most common seem to be aluminide coatings, alloying with copper and addition of steam.

# Pitting Corrosion

Pitting on the Nandu River Iron Bridge

Pitting corrosion, or pitting, is a form of extremely localized corrosion that leads to the creation of small holes in the metal. The driving power for pitting corrosion is the depassivation of a small area, which becomes anodic while an unknown but potentially vast area becomes cathodic, leading to very localised galvanic corrosion. The corrosion penetrates the mass of the metal, with a limited diffusion of ions. The mechanism of pitting corrosion is probably the same as crevice corrosion.

## Mechanism

The more conventional explanation for pitting corrosion is that it is an autocatalytic process. Metal oxidation results in localised acidity that is maintained by the spatial separation of the cathodic and anodic half-reactions, which creates a potential gradient and electromigration of aggressive anions into the pit. For example, when a metal is present in an oxygenated NaCl electrolyte, the pit acts as anode and the metal surface acts as cathode. The localised production of positive metal ions in the pit gives a local excess of positive charge which attracts the negative chlorine ions from the electrolyte to produce charge neutrality. The pit contains a high concentration of MCl molecules which react with water to produce HCl, the metal hydroxide, and H+ ions, accelerating the corrosion process.In the pit, the oxygen concentration is essentially zero and all of the cathodic oxygen reactions take place on the metal surface outside the pit. The pit is anodic and the locus of rapid dissolution of the metal. The metal corrosion initiation is autocatalytic in nature however its propagation is not.

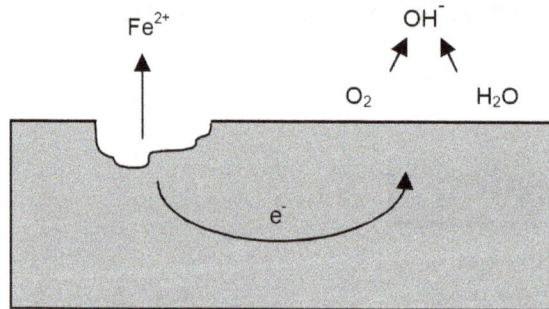

Diagram showing a mechanism of localised corrosion developing
on metal in a solution containing oxygen

This kind of corrosion is extremely insidious, as it causes little loss of material with the small effect on its surface, while it damages the deep structures of the metal. The pits on the surface are often obscured by corrosion products.

Pitting can be initiated by a small surface defect, being a scratch or a local change in composition, or a damage to the protective coating. Polished surfaces display higher resistance to pitting.

## Susceptible Alloy / Environment Combinations

Pitting Corrosion is defined by localised attack (microns - millimetres in diameter) in an otherwise passive surface and only occurs for specific alloy / environment combinations. Thus, this type of corrosion typically occurs in alloys that are protected by a tenacious (passivating) oxide film such as stainless steels, nickel alloys, aluminium alloys in environments that contain an aggressive species such as chlorides (Cl-). In contrast, alloy / environment combinations where the passive film is not very protective usually will not produce pitting corrosion. A good example of the importance of environment / alloy combinations is carbon steel. In environments where the pH is less than 10, carbon steel does not form a passivating oxide film and the addition of chloride results in uniform attack over the entire surface. However, at pH greater than 10 (alkaline) the oxide is very protecting and the addition of chloride results in pitting corrosion.

Besides chlorides, other anions implicated in pitting include thiosulfates ($S_2O_3^{2-}$), fluorides and iodides. Stagnant water conditions favour pitting. Thiosulfates are particularly aggressive species and are formed by partial oxidation of pyrite, or partial reduction of sulfate. Thiosulfates are a concern for corrosion in many industries: sulfide ores processing, oil wells and pipelines transporting soured oils, Kraft paper production plants, photographic industry, methionine and lysine factories.

Corrosion inhibitors, when present in sufficient amount, will provide protection against pitting. However, too low level of them can aggravate pitting by forming local anodes.

## Engineering Failures Due to Pitting Corrosion

A corrosion pit on the outside wall of a pipeline at a coating defect before and after abrasive blasting.

A single pit in a critical point can cause a great deal of damage. One example is the explosion in Guadalajara, Mexico on April 22, 1992, when gasoline fumes accumulated in sewers destroyed kilometres of streets. The vapours originated from a leak of gasoline through a single hole formed by corrosion between a steel gasoline pipe and a zinc-plated water pipe. Firearms can also suffer from pitting, most notably in the bore of the barrel when corrosive ammunition is used and the barrel is not cleaned soon afterwards. Deformities in the bore caused by pitting can greatly reduce the firearm's accuracy. To prevent pitting in firearm bores, most modern firearms have a bore lined with chromium.

Pitting corrosion can also help initiate stress corrosion cracking, as happened when a single eyebar on the Silver Bridge, West Virginia failed and ended up killing 46 people on the bridge in December, 1967.

# Selective Leaching

Selective leaching, also called dealloying, demetalification, parting and selective corrosion, is a corrosion type in some solid solution alloys, when in suitable conditions a component of the alloys is preferentially leached from the material. The less noble metal is removed from the alloy by a microscopic-scale galvanic corrosion mechanism. The most susceptible alloys are the ones containing metals with high distance between each other in the galvanic series, e.g. copper and zinc in brass. The elements most typically undergoing selective removal are zinc, aluminium, iron, cobalt, chromium, and others.

## Leaching of Zinc

The most common example is selective leaching of zinc from brass alloys containing more than 15% zinc (dezincification) in the presence of oxygen and moisture, e.g. from brass taps in chlorine-containing water. It is believed that both copper and zinc gradually dissolve out simultaneously, and copper precipitates back from the solution. The material remaining is a copper-rich sponge with poor mechanical properties, and a color changed from yellow to red. Dezincification can be caused by water containing sulfur, carbon dioxide, and oxygen. Stagnant or low velocity waters tend to promote dezincification.

To combat this, arsenic or tin can be added to brass, or gunmetal can be used instead. Dezincification resistant brass (DZR), also known as Brass C352 is an alloy used to make pipe fittings for use with potable water. Plumbing fittings that are resistant to dezincification are appropriately marked, with the letters "CR" (Corrosion Resistant) or DZR (dezincification resistant) in the UK, and the letters "DR" (dezincification resistant) in Australia.

## Graphitic Corrosion

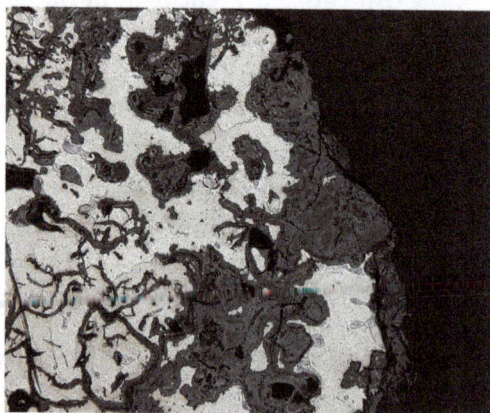

Selective corrosion on cast iron. Magnification 100x

Selective corrosion on cast iron. Magnification 500x

Graphitic corrosion is selective leaching of iron from grey cast iron, where iron gets removed and graphite grains remain intact. Affected surfaces develop a layer of graphite, rust, and metallurgical impurities that may inhibit further leaching. The effect can be substantially reduced by alloying the cast iron with nickel.

## Leaching of other Elements

Dealuminification is a corresponding process for aluminum alloys. Similar effects for different metals are decarburization (removal of carbon from the surface of alloy), decobaltification, denickelification, etc.

## Countermeasures

Countermeasures involve using alloys not susceptible to grain boundary depletion, using a suitable heat treatment, altering the environment (e.g. lowering oxygen content), and/or use cathodic protection.

## Uses

Selective leaching can be used to produce powdered materials with extremely high surface area, such as Raney nickel. Selective leaching can be the pre-final stage of depletion gilding.

## References

- Islander, R.L., Devinny, J.S., Mansfeld, F., Postyn, A., Shih, H., 1991. Microbial ecology of crown corrosion in sewers. Journal of Environmental Engineering 117, 751-770

- Graham-Cumming, John (2009). "Tempio Voltiano". The Geek Atlas: 128 Places Where Science and Technology Come Alive. O'Reilly Media. p. 97. ISBN 9780596523206

- Ankersmit, Bart; Griesser-Stermscheg, Martina; Selwyn, Lindsie; Sutherland, Susanne. ""Rust Never Sleeps: Recognizing Metals and Their Corrosion Products"" (PDF). depotwijzer. Parks Canada. Retrieved 23 July 2016

- Gräfen, H.; Horn, E. M.; Schlecker, H.; Schindler, H. (2000). "Corrosion". Ullmann's Encyclopedia of Industrial Chemistry. Wiley-VCH. doi:10.1002/14356007.b01_08

- Morton R.L., Yanko W.A., Grahom D.W., Arnold R.G. (1991) Relationship between metal concentrations and crown corrosion in Los Angeles County sewers. Research Journal of Water Pollution Control Federation, 63, 789–798

- Sawyer, Clair N. & McCarty, Perry L. Chemistry for Sanitary Engineers (2nd edition) McGraw-Hill (1967) ISBN 0-07-054970-2

- Mullan, Jeff (April 6, 2011). "Tunnel Safety Ceiling Light Fixture Update" (PDF). Report to the MassDOT Board of Directors. MassDOT. Retrieved 2012-04-09.

- G. Butler and J. G. Beynon (1967): "The corrosion of mild steel in boiling salt solutions". Corrosion Science 7, pages 385-404. doi:10.1016/S0010-938X(67)80052-0

- Fairand, B. P. (1972). "Laser Shock-Induced Microstructural and Mechanical Property Changes in 7075 Aluminum". Journal of Applied Physics. 43 (9): 3893. doi:10.1063/1.1661837

- Weismann, D. & Lohse, M. (Hrsg.): "Sulfid-Praxishandbuch der Abwassertechnik; Geruch, Gefahr, Korrosion verhindern und Kosten beherrschen!" 1. Auflage, VULKAN-Verlag, 2007, ISBN 978-3-8027-2845-7

- Anderson, Colin. "Protection of Ships Against Corrosion" (PDF). Protection of Ships Lecture Series. Newcastle University (UK). Retrieved 12 December 2012

- Banks, Bruce A.; De Groh, Kim K.; Rutledge, Sharon K.; Haytas, Christy A. (1999). "Consequences of atomic oxygen interaction with silicone and silicone contamination on surfaces in low earth orbit". Proc. SPIE. 3784: 62. Bibcode:1999SPIE.3784...62B. doi:10.1117/12.366725

- Fabbro, R.; Fournier, J.; Ballard, P.; Devaux, D.; Virmont, J. (1990). "Physical Study of Laser Produced Plasma in Confined Geometry". Journal of Applied Physics. 68 (2): 775. doi:10.1063/1.346783

- Carl Branan Rules of thumb for chemical engineers: a manual of quick, accurate solutions to everyday process engineering problems Gulf Professional Publishing, 2005, ISBN 0-7506-7856-9 p. 294

- "About Corrosion and Ballast Water Treatment Systems" (PDF). OceanSaver Ballast and Corrosion Control. Retrieved 12 December 2012

- Metz, S. A. (1971). "Production of Vacancies by Laser Bombardment". Applied Physics Letters. 19 (6): 207. doi:10.1063/1.1653886

# Corrosion: Effects and Mechanism

Corrosion is an unwanted effect that occurs in metals. It causes degradation of materials as well as other forms of wear and tear. Tribocorrosion, fretting, stress corrosion cracking are some of the topics explained here. This chapter elucidates the crucial theories and principles of corrosion engineering.

## Stress Corrosion Cracking

A close-up of the surface of a steel pipeline showing indications of stress corrosion cracking (two clusters of small black lines) revealed by magnetic particle inspection.
Cracks which would normally have been invisible are detectable due to the magnetic particles clustering at the crack openings. The scale at the bottom is in millimetres.

Stress corrosion cracking (SCC) is the growth of crack formation in a corrosive environment. It can lead to unexpected sudden failure of normally ductile metals subjected to a tensile stress, especially at elevated temperature. SCC is highly chemically specific in that certain alloys are likely to undergo SCC only when exposed to a small number of chemical environments. The chemical environment that causes SCC for a given alloy is often one which is only mildly corrosive to the metal otherwise. Hence, metal parts with severe SCC can appear bright and shiny, while being filled with microscopic cracks. This factor makes it common for SCC to go undetected prior to failure. SCC often progresses rapidly, and is more common among alloys than pure metals. The specific environment is of crucial importance, and only very small concentrations of certain highly active chemicals are needed to produce catastrophic cracking, often leading to devastating and unexpected failure.

The stresses can be the result of the crevice loads due to stress concentration, or can be caused by the type of assembly or residual stresses from fabrication (e.g. cold working); the residual stresses can be relieved by annealing or other surface treatments.

## Metals Attacked

Certain austenitic stainless steels and aluminium alloys crack in the presence of chlorides, mild steel cracks in the presence of alkali (boiler cracking) and nitrates, copper alloys crack in ammoniacal solutions (season cracking). This limits the usefulness of austenitic stainless steel for containing water with higher than few ppm content of chlorides at temperatures above 50 °C. Worse still, high-tensile structural steels crack in an unexpectedly brittle manner in a whole variety of aqueous environments, especially containing chlorides. With the possible exception of the latter, which is a special example of hydrogen cracking, all the others display the phenomenon of subcritical crack growth, i.e. small surface flaws propagate (usually smoothly) under conditions where fracture mechanics predicts that failure should not occur. That is, in the presence of a corrodent, cracks develop and propagate well below $K_{Ic}$. In fact, the subcritical value of the stress intensity, designated as $K_{Iscc}$, may be less than 1% of $K_{Ic}$, as the following table shows:

| Alloy | $K_{Ic}$<br>MN/m$^{3/2}$ | SCC environment | $K_{Iscc}$<br>MN/m$^{3/2}$ |
|---|---|---|---|
| 13Cr steel | 60 | 3% NaCl | 12 |
| 18Cr-8Ni | 200 | 42% MgCl$_2$ | 10 |
| Cu-30Zn | 200 | NH$_4$OH, pH7 | 1 |
| Al-3Mg-7Zn | 25 | Aqueous halides | 5 |
| Ti-6Al-1V | 60 | 0.6M KCl | 20 |

## Polymers Attacked

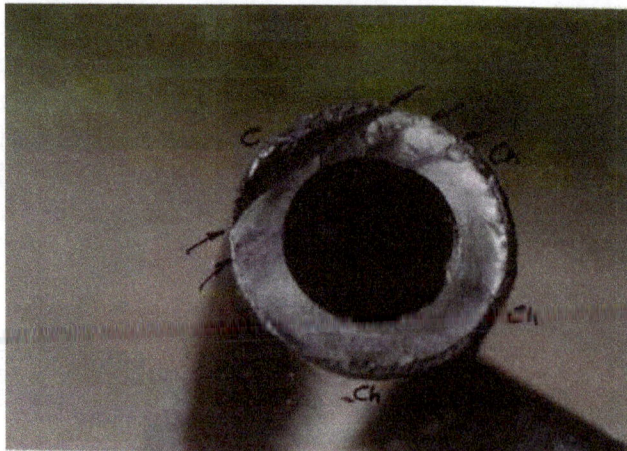

Close-up of broken nylon fuel pipe connector caused by SCC

A similar process (environmental stress cracking) occurs in polymers, when products are exposed to specific solvents or aggressive chemicals such as acids and alkalis. As with metals, attack is confined to specific polymers and particular chemicals. Thus polycarbonate is sensitive to attack by alkalis, but not by acids. On the other hand, polyesters are readily degraded by acids, and SCC is a likely failure mechanism. Polymers are susceptible to environmental stress cracking where attacking agents do not necessarily degrade the materials chemically. Nylon is sensitive to degradation by acids, a process known as hydrolysis, and nylon mouldings will crack when attacked by strong acids.

For example, the fracture surface of a fuel connector showed the progressive growth of the crack from acid attack (Ch) to the final cusp (C) of polymer. In this case the failure was caused by hydrolysis of the polymer by contact with sulfuric acid leaking from a car battery. The degradation reaction is the reverse of the synthesis reaction of the polymer:

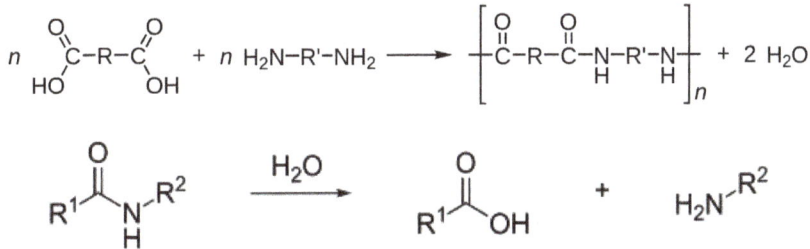

Cracks can be formed in many different elastomers by ozone attack, another form of SCC in polymers. Tiny traces of the gas in the air will attack double bonds in rubber chains, with natural rubber, styrene-butadiene rubber, and nitrile butadiene rubber being most sensitive to degradation. Ozone cracks form in products under tension, but the critical strain is very small. The cracks are always oriented at right angles to the strain axis, so will form around the circumference in a rubber tube bent over. Such cracks are very dangerous when they occur in fuel pipes because the cracks will grow from the outside exposed surfaces into the bore of the pipe, so fuel leakage and fire may follow. The problem of ozone cracking can be prevented by adding anti-ozonants to the rubber before vulcanization. Ozone cracks were commonly seen in automobile tire sidewalls, but are now seen rarely thanks to the use of these additives. On the other hand, the problem does recur in unprotected products such as rubber tubing and seals.

## Ceramics Attacked

This effect is significantly less common in ceramics which are typically more resilient to chemical attack. Although phase changes are common in ceramics under stress these usually result in toughening rather than failure. Recently studies have shown that the same driving force for this toughening mechanism can also enhance oxidation of reduced cerium oxide resulting in slow crack growth and spontaneous failure of dense ceramic bodies.

## Crack Growth

The subcritical nature of propagation may be attributed to the chemical energy released as the crack propagates. That is,

*elastic energy released + chemical energy = surface energy + deformation energy*

The crack initiates at $K_{Iscc}$ and thereafter propagates at a rate governed by the slowest process, which most of the time is the rate at which corrosive ions can diffuse to the crack tip. As the crack advances so $K$ rises (because crack length appears in the calculation of stress intensity). Finally it reaches $K_{Ic}$, whereupon fast fracture ensues and the component fails. One of the practical difficulties with SCC is its unexpected nature. Stainless steels, for example, are employed because under most conditions they are "passive", i.e. effectively inert. Very often one finds a single crack has propagated while the rest of the metal surface stays apparently unaffected. The crack propagates perpendicular to the applied stress.

## Prevention

SCC is the result of a combination of three factors – a susceptible material, exposure to a corrosive environment, and tensile stresses above a threshold. If any one of these factors are eliminated, SCC initiation becomes impossible. There are, consequently, a number of approaches that can be used to prevent or at least delay the onset of SCC. In an ideal world a stress corrosion cracking control strategy will start operating at the design stage, and will focus on the selection of material, the limitation of stress and the control of the environment. The skill of the engineer then lies in selecting the strategy that delivers the required performance at minimum cost. In this context it should be noted that part of the performance requirements relate to the acceptability of failure. The primary containment pressure vessel in a nuclear reactor obviously requires a very low risk of failure. For the pressed brass decorative trim on a light switch, the occasional stress corrosion crack is not going to be a serious problem, although frequent failures would have an undesirable impact on product returns and the manufacturer's image. The conventional approach to controlling the problem has been to develop new alloys that are more resistant to SCC. This is a costly proposition and can require a massive time investment to achieve only marginal success.

## Selection and Control of Material

The first line of defence in controlling stress corrosion cracking is to be aware of the possibility at the design and construction stages. By choosing a material that is not susceptible to SCC in the service environment, and by processing and fabricating it correctly, subsequent SCC problems can be avoided. Unfortunately, it is not always quite that simple. Some environments, such as high temperature water, are very aggressive, and will cause SCC of most materials. Mechanical requirements, such as a high yield strength, can be very difficult to reconcile with SCC resistance (especially where hydrogen embrittlement is involved).

## Control of Stress

As one of the requirements for stress corrosion cracking is the presence of stress in the components, one method of control is to eliminate that stress, or at least reduce it below the threshold stress for SCC. This is not usually feasible for working stresses (the stress that the component is intended to support), but it may be possible where the stress causing cracking is a residual stress introduced during welding or forming. Residual stresses can be relieved by stress-relief annealing, and this is widely used for carbon steels. These have the advantage of a relatively high threshold stress for most environments, consequently it is relatively easy to reduce the residual stresses to a low enough level. In contrast austenitic stainless steels have a very low threshold stress for chloride SCC. This, combined with the high annealing temperatures that are necessary to avoid other problems, such as sensitization and sigma phase embrittlement, means that stress relief is rarely successful as a method of controlling SCC for this system. For large structures, for which full stress-relief annealing is difficult or impossible, partial stress relief around welds and other critical areas may be of value. However, this must be done in a controlled way to avoid creating new regions of high residual stress, and expert advice is advisable if this approach is adopted. Stresses can also be relieved mechanically. For example, hydrostatic testing beyond yield will tend to 'even-out' the stresses and thereby reduce the peak residual stress. Similarly laser peening, shot-peening, or grit-blasting tend to introduce a surface compressive stress, and are beneficial for the control of SCC. The uniformity with which these processes are applied is important. If, for example, only the weld region is shot-peened, damaging tensile stresses may be created at the border of the peened area. The compressive residual stresses imparted by laser peening are precisely controlled both in location and intensity, and can be applied to mitigate sharp transitions into tensile regions. Laser peening imparts deep compressive residual stresses on the order of 10 to 20 times deeper than conventional shot peening making it significantly more beneficial at preventing SCC. Laser peening is widely used in the aerospace and power generation industries in gas fired turbine engines.

## Control of Environment

The most direct way of controlling SCC through control of the environment is to remove or replace the component of the environment that is responsible for the problem, though this is not usually feasible. Where the species responsible for cracking are required components of the environment, environmental control options consist of adding inhibitors, modifying the electrode potential of the metal, or isolating the metal from the environment with coatings.

For example, chloride stress corrosion cracking of austenitic stainless steel has been experienced in hot-water jacketed pipes carrying molten chocolate in the food industry. It is difficult to control the temperature, while changing pipe material or eliminating residual stresses associated with welding and forming the pipework is costly and incurs plant downtime. However, this is a rare case where environment may be modified: an ion exchange process may be used to remove chlorides from the heating water.

## Testing of Susceptible Materials

One of the primary methods used to detect and remove materials that are susceptible to SCC is corrosion testing. A variety of SCC corrosion tests exist for different metal alloy.

## Examples

A classic example of SCC is season cracking of brass cartridge cases, a problem experienced by the British army in India in the early 19th century. It was initiated by ammonia from dung and horse manure decomposing at the higher temperatures of the spring and summer. There was substantial residual stress in the cartridge shells as a result of cold forming. The problem was solved by annealing the shells to ameliorate the stress.

A 32-inch diameter gas transmission pipeline, north of Natchitoches, Louisiana, belonging to the Tennessee Gas Pipeline exploded and burned from SCC on March 4, 1965, killing 17 people. At least 9 others were injured, and 7 homes 450 feet from the rupture were destroyed.

SCC caused the catastrophic collapse of the Silver Bridge in December 1967, when an eyebar suspension bridge across the Ohio river at Point Pleasant, West Virginia, suddenly failed. The main chain joint failed and the whole structure fell into the river, killing 46 people in vehicles on the bridge at the time. Rust in the eyebar joint had caused a stress corrosion crack, which went critical as a result of high bridge loading and low temperature. The failure was exacerbated by a high level of residual stress in the eyebar. The disaster led to a nationwide reappraisal of bridges.

Suspended ceilings in indoor swimming pools are safety-relevant components. As was demonstrated by the collapses of the ceiling of the Uster (Switzerland) indoor swimming pool (1985) and again at Steenwijk (Netherlands, 2001), attention must be paid to selecting suitable materials and inspecting the state of such components. The reason for the failures was stress corrosion cracking of metal fastening components made of stainless steel. The active chemical was chlorine added to the water as a disinfectant.

## Season Cracking

Season cracking is a form of stress-corrosion cracking of brass cartridge cases originally reported from British forces in India. During the monsoon season, military activity was temporarily reduced, and ammunition was stored in stables until the dry weather returned. Many brass cartridges were subsequently found to be cracked, especially where the case was crimped to the bullet. It was not until 1921 that the phenomenon was explained by Moor, Beckinsale and Mallinson: ammonia from horse urine, combined with the residual stress in the cold-drawn metal of the cartridges, was responsible for the cracking.

Cracking in brass caused by ammonia attack

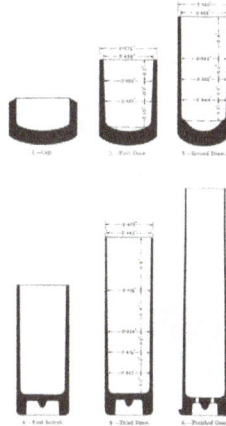

Different draw ratios for brass cartridge case

Season cracking is characterised by deep brittle cracks which penetrate into affected components. If the cracks reach a critical size, the component can suddenly fracture, sometimes with disastrous results. However, if the concentration of ammonia is very high, then attack is much more severe, and attack over all exposed surfaces occurs. The problem was solved by annealing the brass cases after forming so as to relieve the residual stresses.

## Ammonia

Attack takes the form of a reaction between ammonia and copper to form the cuprammonium ion, formula $[Cu(NH_3)_4]^{2+}$, a chemical complex which is water-soluble, and hence washed from the growing cracks. So the problem of cracking can also occur in copper and any other copper alloy, such as bronze. The tendency of copper to react with ammonia was exploited in making rayon, and the deep blue colour of the aqueous solution of copper(II) oxide in ammonia is known as Schweizer's reagent.

## Materials

Although the problem was first found in brass, any alloy containing copper will be susceptible to the problem. It includes copper itself (as used in pipe for example), bronzes and other alloys with a significant copper content.

## Tribocorrosion

Tribocorrosion is a material degradation process due to the combined effect of corrosion and wear. The name tribocorrosion expresses the underlying disciplines of tribology and corrosion. Tribology is concerned with the study of friction, lubrication and wear (its name comes from the Greek "tribo" meaning to rub) and corrosion is concerned with the chemical and electrochemical interactions between a material, normally a metal, and its environment. As a field of research tribocorrosion is relatively new, but tribocorrosion phenomena have been around ever since machines and installations are being used.

Wear is a mechanical material degradation process occurring on rubbing or impacting surfaces, while corrosion involves chemical or electrochemical reactions of the material. Corrosion may accelerate wear and wear may accelerate corrosion. One then speaks of corrosion accelerated wear or wear accelerated corrosion. Both these phenomena, as well as fretting corrosion (which results from small amplitude oscillations between contacting surfaces) fall into the broader category of tribocorrosion. Erosion-corrosion is another tribocorrosion phenomenon involving mechanical and chemical effects: impacting particles or fluids erode a solid surface by abrasion, chipping or fatigue while simultaneously the surface corrodes.

## Phenomena in Different Engineering Fields

Tribocorrosion occurs in many engineering fields. It reduces the life-time of pipes, valves and pumps, of waste incinerators, of mining equipment or of medical implants, and it can affect the safety of nuclear reactors or of transport systems. On the other hand, tribocorrosion phenomena can also be applied to good use, for example in the chemical-mechanical planarization of wafers in the electronics industry or in metal grinding and cutting in presence of aqueous emulsions. Keeping this in mind, we may define tribocorrosion in a more general way independently of the notion of usefulness or damage or of the particular type of mechanical interaction: Tribocorrosion concerns the irreversible transformation of materials or of their function as a result of simultaneous mechanical and chemical/electrochemical interactions between surfaces in relative motion.

## Biotribocorrosion

Biotribocorrosion covers the science of surface transformations resulting from the interactions of mechanical loading and chemical/electrochemical reactions that occur between elements of a tribological system exposed to biological environments. It has been studied for aritificial joint prostheses. It is important to understand material degradation processes for joint implants to achieve longer service life and better safety issues for such devices.

## Passive Metals

While tribocorrosion phenomena may affect many materials, they are most critical for metals, especially the normally corrosion resistant so-called passive metals. The vast majority of corrosion resistant metals and alloys used in engineering (stainless steels, titanium, aluminium etc.) fall into this category. These metals are thermodynamically unstable in the presence of oxygen or water and they derive their corrosion resistance from the presence at the surface of a thin oxide film, called the passive film, which acts as a protective barrier between the metal and its environment. Passive films are usually just a few atomic layers thick. Nevertheless, they can provide excellent corrosion protection because if damaged accidentally they spontaneously self-heal by metal oxidation. However, when a metal surface is subjected to severe rubbing or to a stream of impacting particles the passive film damage becomes continuous and extensive. The self-healing process may no longer be effective and in addition it requires a high rate of metal oxidation. In other words, the underlying metal will strongly corrode before the protective passive film is reformed, if at all. In such a case, the total material loss due to tribocorrosion will be much higher than the sum of wear and corrosion one would measure in experiments with the same metal where only wear or only corrosion takes place. The example illustrates the fact that the rate of tribocorrosion is not simply the addition of the rate of wear and the rate of corrosion but it is strongly affected by synergistic and antagonistic effects between mechanical and chemical mechanisms. To study such effects in the laboratory, one most often uses mechanical wear testing rigs which are equipped with an electrochemical cell. This permits one to control independently the mechanical and chemical parameters. For example, by imposing a given potential to the rubbing metal one can simulate the oxidation potential of the environment and in addition, under certain conditions, the current flow is a measure of the instantaneous corrosion rate. Volume loss due to electrochemical dissolution can be measured by Faraday's laws of electrolysis and subtracted from total volume loss in tribocorrosion so the sum of mechanical wear loss and the synergies can be calculated. For a deeper understanding tribocorrosion experiments are supplemented by detailed microscopic and analytical studies of the contacting surfaces.

At high temperatures, the more rapid generation of oxide due to a combination of temperature and tribological action during sliding wear can generate potentially wear resistant oxide layers known as 'glazes'. Under such circumstances, tribocorrosion can be used potentially in a beneficial way.

## Fretting

Fretting refers to wear and sometimes corrosion damage at the asperities of contact surfaces. This damage is induced under load and in the presence of repeated relative surface motion, as induced for example by vibration. The ASM Handbook on Fatigue and Fracture defines fretting as: "*A special wear process that occurs at the contact area between two materials under load and subject to minute relative motion by vibration or some other force.*" Fretting tangibly downgrades the surface layer quality producing increased surface roughness and micropits; which reduces the fatigue strength of the components.

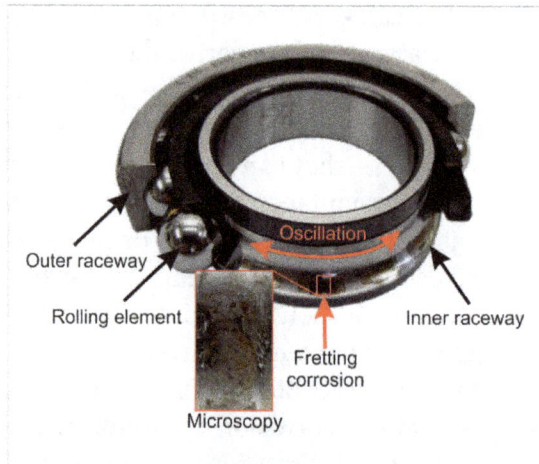

Fretting corrosion on the inner raceway of a ball bearing

The amplitude of the relative sliding motion is often in the order from micrometers to millimeters, but can be as low as 3 nanometers.

The contact movement causes mechanical wear and material transfer at the surface, often followed by oxidation of both the metallic debris and the freshly exposed metallic surfaces. Because the oxidized debris is usually much harder than the surfaces from which it came, it often acts as an abrasive agent that increases the rate of fretting.

The distinction between false brinelling and fretting corrosion has been extensively discussed in the literature.. The main difference is that false brinelling occurs under lubricated and fretting under dry contact conditions. Between false brinelling and fretting corrosion also exists a time-dependend connection.

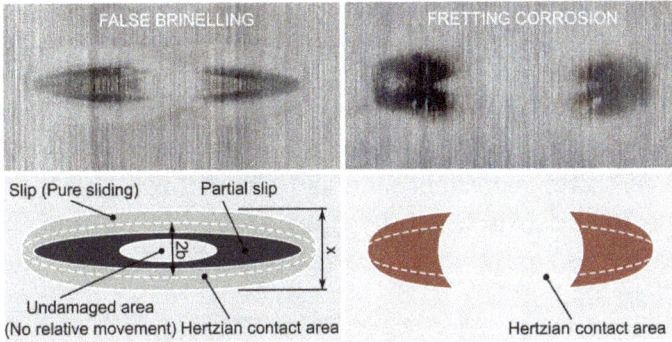

Different areas of typical false brinelling and fretting corrosion damage in a ball bearing

## Steel

Fretting damage in steel can be identified by the presence of a pitted surface and fine 'red' iron oxide dust resembling cocoa powder. Strictly this debris is not 'rust' as its production requires no water. The particles are much harder than the steel surfaces in contact, so abrasive wear is inevitable; however, particulates are not required to initiate fret.

## Products Affected

Fretting examples include wear of drive splines on driveshafts, wheels at the lug bolt inter-face, and cylinder head gaskets subject to differentials in thermal expansion coefficients.

There is currently a focus on fretting research in the aerospace industry. The dovetail blade-root connection and the spline coupling of gas turbine aero engines experience fretting.

Fretting corrosion can also occur in variable speed blower fan motors installed in gas or electric heaters. Affected motors are ECM 3.0 blower motors with a 4 pin communi-cation connector.

Another example in which fretting corrosion may occur are the pitch bearings of mod-ern wind turbines, which operate under oscillation motion to control the power and loads of the turbine.

## Fretting Fatigue

Fretting decreases fatigue strength of materials operating under cycling stress. This can re-sult in *fretting fatigue*, whereby fatigue cracks can initiate in the fretting zone. Afterwards, the crack propagates into the material. Lap joints, common on airframe surfaces, are a prime location for fretting corrosion. This is also known as frettage or fretting corrosion.

## Mitigation

The fundamental way to prevent fretting is to design for no relative motion of the sur-faces at the contact. Surface roughness plays an important role as fretting normally occurs by the contact of the asperities of the mating surfaces. Lubricants are often employed to mitigate fretting because they reduce friction and inhibit oxidation.

Soft materials often exhibit higher susceptibility to fretting than hard materials of a similar type. The hardness ratio of the two sliding materials also has an effect on fretting wear. However, softer materials such as polymers can show the opposite effect when they capture hard debris which becomes embedded in their bearing surfaces. They then act as a very effective abrasive agent, wearing down the harder metal with which they are in contact.

# Erosion Corrosion of Copper Water Tubes

Erosion corrosion, also known as impingement damage, is the combined effect of corrosion and erosion caused by rapid flowing turbulent water. It is probably the second most common cause of copper tube failures behind Type 1 pitting which is also known as Cold Water Pitting of Copper Tube.

Copper Water Tubes Copper tubes have been used to distribute drinking water within buildings for many years, and hundreds of miles are installed throughout Europe every year. The long life of copper when exposed to natural waters is a result of its thermodynamic stability, its high resistance to reacting with the environment, and the formation of insoluble corrosion products that insulate the metal from the environment. The corrosion rate of copper in most drinkable waters is less than 2.5 μm/year, at this rate a 15 mm tube with a wall thickness of 0.7 mm would last for about 280 years. In some soft waters the general corrosion rate may increase to 12.5 μm/year, but even at this rate it would take over 50 years to perforate the same tube.

## Occurrence

If the general water speed or the degree of local turbulence in an installation is high, the protective film that would normally be formed on a copper tube as a result of slight initial corrosion, may be torn off the surface locally, permitting further corrosion to take place at that point. If this process continues it can produce deep localised attack of the type known as erosion-corrosion or impingement damage. The actual attack on the metal is by the corrosive action of the water to which it is exposed while the erosive factor is the mechanical removal of the corrosion product from the surface.

Impingement attack produces highly characteristic water-swept pits, which are often horseshoe shaped, or it can produce broader areas of attack. The leading edge of the pit is frequently undercut by the swirling action of the water. Usually, the surface of the metal within the pits or areas of attack is smooth and carries no substantial corrosion product. Erosion-corrosion is known to occur in pumped-circulation hot water distribution systems, and even in cold water distribution systems, if the water velocities are too high. The factors influencing the attack include the chemical character of the water passing through the system, the temperature, the average water velocity in the system and the presence of any local features likely to induce turbulence in the water stream.

It is unusual for the general water velocity in a system to be so high that impingement attack occurs throughout the whole of the copper pipework. More commonly, the velocity is just sufficiently low for satisfactory protective films to be formed and to remain in position on most of the system, with impingement damage more likely to occur where there is an abrupt change in the direction of water flow giving rise to a high degree of turbulence, such as at tee pieces and elbow fittings. It is not generally realised how great an effect small obstructions can have on the flow pattern of water in a pipe-work system and the extent to which they can induce turbulence and cause corrosion-erosion. For example, it is most important, as far as possible, to ensure that copper tubes cut with a tube cutter are deburred before making the joint. Also a gap between the tube end and the stop in the fitting, due to the tube not having been cut to the correct length and fully inserted into the socket of the fitting, can also induce turbulence in the water stream.

## Recommendations

The rate of impingement attack on copper also depends to some extent on the temperature of the water. The maximum velocities for fresh waters at different temperatures recommended in Sweden are given in the table below. These figures are for aerated waters of pH not less than about 7.

Recommended Maximum Water Velocities at Different Temperatures for Copper (m/s)

|  | 10 °C | 50 °C | 70 °C | 90 °C |
|---|---|---|---|---|
| For pipes that can be replaced: | 4.0 | 3.0 | 2.5 | 2.0 |
| For pipes that cannot be replaced: | 2.0 | 1.5 | 1.3 | 1.0 |
| For short connections to taps, etc.: | 16.0 | 12.0 | 10.0 | 8.0 |

These velocities give a risk of impingement attack and are acceptable only for small bore connections to taps, flushing cisterns etc., through which water flow is intermittent.

BS 6700 gives the following maximum water velocities although it does note that these are currently under investigation and the velocities specified will be amended if the results of this investigation so require.

| Water Temperature °C | Maximum Water Velocity (m/s) |
|:---:|:---:|
| 10 | 3.0 |
| 50 | 3.0 |
| 70 | 2.5 |
| 90 | 2.0 |

The minimum water speed at which copper pipes suffer impingement attack depends also to some extent on water composition. Aggressive waters that tend to be cupro-solvent are the most likely to give rise to impingement attack. Installations in large buildings where flow rates may be high and water is in continuous circulation are much more susceptible to attack than ordinary domestic installations. A high mineral content or a pH below 7 is likely to increase the possibility of corrosion-erosion occurring while a positive Langelier Index and consequent tendency to deposit a calcium carbonate scale is generally beneficial. The presence or absence of colloidal organic matter is also probably of some importance.

Remedial measures for impingement attack include modifications to the system to re-duce the average water velocity, e.g. by using larger diameter tubes or, if appropriate, to lower the pump speed, and/or to redesign the part of the installation concerned to eliminate the cause of local turbulence, e.g. by using slow or swept bends and tee fit-tings rather than elbows and square tees. It is important to minimise the possibility of any local turbulence occurring by ensuring that the ends of tubes cut with a tube cutter are deburred and that the tubes are inserted fully to the stops in the fitting before the joints are made, as referred to earlier in this chapter. In some cases, where the above approaches are not possible, the length of copper tube affected can sometimes be re-placed by materials more resistant to corrosion-erosion, e.g. 90/10 copper-nickel (BS Designation CN102) using appropriate fittings, or stainless steel to BS 4127:1994.

# Corrosion Fatigue

Corrosion fatigue is fatigue in a corrosive environment. It is the mechanical degra-dation of a material under the joint action of corrosion and cyclic loading. Nearly all engineering structures experience some form of alternating stress, and are exposed to harmful environments during their service life. The environment plays a signif-icant role in the fatigue of high-strength structural materials like steel, aluminum alloys and titanium alloys. Materials with high specific strength are being devel-oped to meet the requirements of advancing technology. However, their usefulness depends to a large extent on the degree to which they resist corrosion fatigue. The effects of corrosive environments on the fatigue behavior of metals were studied as early as 1930. The phenomenon should not be confused with stress corrosion cracking, where corrosion (such as pitting) leads to the development of brittle cracks, growth and failure. The only requirement for corrosion fatigue is that the sample be under tensile stress.

## Effect of Corrosion on S-N Diagram

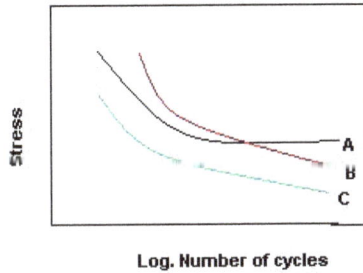

Effect of corrosion on S-N curve
A- Air, B- Corrosive 1, C- Corrosive 2

Effect of corrosion on S-N diagram

The effect of corrosion on a smooth-specimen S-N diagram is shown schematically on the right. Curve A shows the fatigue behavior of a material tested in air. A fatigue threshold (or limit) is seen in curve A, corresponding to the horizontal part of the curve. Curves B and C represent the fatigue behavior of the same material in two corrosive environments. In curve B, the fatigue failure at high stress levels is retarded, and the fatigue limit is eliminated. In curve C, the whole curve is shifted to the left; this indicates a general lowering in fatigue strength, accelerated initiation at higher stresses and elimination of the fatigue limit. To meet the needs of advancing technology, higher-strength materials are developed through heat treatment or alloying. Such high-strength materials generally exhibit higher fatigue limits, and can be used at higher service stress levels even under fatigue loading. However, the presence of a corrosive environment during fatigue loading eliminates this stress advantage, since the fatigue limit becomes almost insensitive to the strength level for a particular group of alloys. This effect is schematically shown for several steels in the diagram on the left, which illustrates the debilitating effect of a corrosive environment on the functionality of high-strength materials under fatigue.

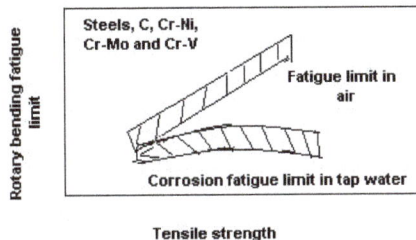

Effect of corrosion on fatigue limit of steels of
increasing strength levels

Effect of corrosion on fatigue limits of steels

Corrosion fatigue in aqueous media is an electrochemical behavior. Fractures are initiated either by pitting or persistent slip bands. Corrosion fatigue may be reduced by alloy additions, inhibition and cathodic protection, all of which reduce pitting. Since corrosion-fatigue cracks initiate at a metal's surface, surface treatments like plating, cladding, nitriding and shot peening were found to improve the materials' resistance to this phenomenon.

## Crack-propagation Studies in Corrosion Fatigue

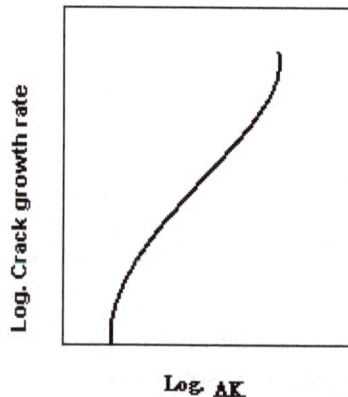

Sub critical crack growth in fatigue

Schematic of typical fatigue-crack-growth behavior

In normal fatigue-testing of smooth specimens, about 90 percent is spent in crack nucleation and only the remaining 10 percent in crack propagation. However, in corrosion fatigue crack nucleation is facilitated by corrosion; typically, about 10 percent of life is sufficient for this stage. The rest (90 percent) of life is spent in crack propagation. Thus, it is more useful to evaluate crack-propagation behavior during corrosion fatigue. Fracture mechanics uses pre-cracked specimens, effectively measuring crack-propagation behavior. For this reason, emphasis is given to crack-propagation velocity measurements (using fracture mechanics) to study corrosion fatigue. Since fatigue crack grows in a stable fashion below the critical stress-intensity factor for fracture (fracture toughness), the process is called sub-critical crack growth.

In this log-log plot, the crack-propagation velocity is plotted against the applied stress-intensity range. Generally there is a threshold stress-intensity range, below which crack-propagation velocity is insignificant. Three stages may be visualized in this plot. Near the threshold, crack-propagation velocity increases with increasing stress-intensity range. In the second region, the curve is nearly linear and follows Paris' law(6); in the third region crack-propagation velocity increases rapidly, with the stress-intensity range leading to fracture at the fracture-toughness value. Crack propagation under corrosion fatigue may be classified as a) true corrosion fatigue, b) stress corrosion fatigue or c) a combination of true, stress and corrosion fatigue.

# True Corrosion Fatigue

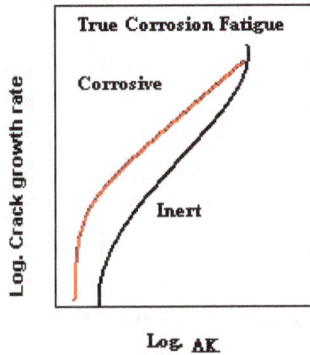

Crack-growth behavior under true corrosion fatigue

In true corrosion fatigue, the fatigue-crack-growth rate is enhanced by corrosion; this effect is seen in all three regions of the fatigue-crack growth-rate diagram. The diagram on the left is a schematic of crack-growth rate under true corrosion fatigue; the curve shifts to a lower stress-intensity-factor range in the corrosive environment. The threshold is lower (and the crack-growth velocities higher) at all stress-intensity factors. Specimen fracture occurs when the stress-intensity-factor range is equal to the applicable threshold-stress-intensity factor for stress-corrosion cracking. When attempting to analyze the effects of corrosion fatigue on crack growth in a particular, both corrosion type and fatigue load levels affect crack growth in varying degrees. Common types of corrosion include filiform, pitting, exfoliation, intergranular; each will affect crack growth in a particular material in a distinct way. For instance, pitting will often be the most damaging type of corrosion, degrading a material's performance (by increasing the crack-growth rate) more than any other kind of corrosion; even pits of the order of a material's grain size may substantially degrade a material. The degree to which corrosion affects crack-growth rates also depends on fatigue-load levels; for instance, corrosion can cause a greater increase in crack-growth rates at a low loads than it does at a high load.

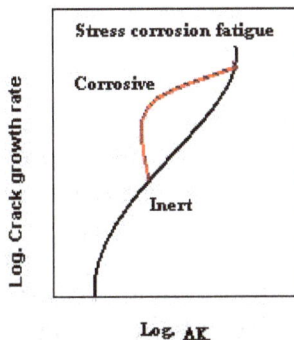

Crack-growth behavior under stress-corrosion fatigue

## Stress-corrosion Fatigue

In materials where the maximum applied-stress-intensity factor exceeds the stress-corrosion cracking-threshold value, stress corrosion adds to crack-growth velocity. This is shown in the schematic on the right. In a corrosive environment, the crack grows due to cyclic loading at a lower stress-intensity range; above the threshold stress intensity for stress corrosion cracking, additional crack growth (the red line) occurs due to SCC. The lower stress-intensity regions are not affected, and the threshold stress-intensity range for fatigue-crack propagation is unchanged in the corrosive environment. In the most-general case, corrosion-fatigue crack growth may exhibit both of the above effects; crack-growth behavior is represented in the schematic on the left.

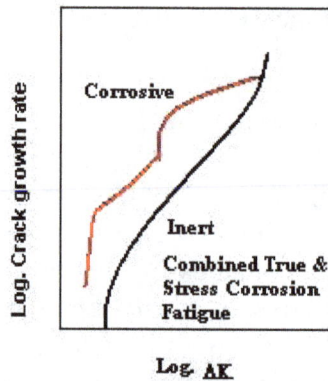

Combined true-corrosion and stress-corrosion fatigue

## Flow-accelerated Corrosion

Flow-accelerated corrosion (FAC), also known as flow-assisted corrosion, is a corrosion mechanism in which a normally protective oxide layer on a metal surface dissolves in a fast flowing water. The underlying metal corrodes to re-create the oxide, and thus the metal loss continues.

By definition, the rate of FAC depends on the flow velocity. FAC often affects carbon steel piping carrying ultra-pure, deoxygenated water or wet steam. Stainless steel does not suffer from FAC. FAC of carbon steel halts in the presence of small amount of oxygen dissolved in water. FAC rates rapidly decrease with increasing water pH.

FAC has to be distinguished from erosion corrosion because the fundamental mechanisms for the two corrosion modes are different. FAC does not involve impingement of particles, bubbles, or cavitation which cause the mechanical (often crater-like) wear on the surface. By contrast to mechanical erosion, FAC involves dissolution of normally poorly soluble oxide by combined electrochemical, water chemistry and mass-transfer phenomena. Nevertheless, the terms FAC and erosion are sometimes used interchangeably because the actual mechanism may, in some cases, be unclear.

FAC was the cause of several high-profile accidents in power plants, for example, a rupture of a high-pressure condensate line in Virginia Power's Surry nuclear plant.

# Oxide Jacking

The expansive force of rusting, which may be called oxide jacking or rust burst, is a phenomenon that can cause damage to structures made of stone, masonry, concrete or ceramics, and reinforced with metal components. A definition is "the displacement of building elements due to the expansion of iron and steel products as the metal rusts and becomes iron oxide". Corrosion of other metals such as aluminum can also cause oxide jacking.

## Physical Process

According to metallurgist Jack Harris, "Oxidation is usually accompanied by a net expansion so that when it occurs in a confined space stresses are generated in the metal component itself or in any surrounding medium such as stone or cement. So much energy is released by oxidation that the stresses generated are of sufficient magnitude to deform or fracture all known materials."

As early as 1915, it was recognized that certain modern metal alloys are more susceptible to excessive oxidation when subjected to weathering than other metals. At that time, there was a trend to replace wrought iron fasteners with mild steel equivalents, which were less expensive. Unexpectedly, the mild steel fasteners failed in real world use much more quickly than anticipated, leading to a return to use of wrought iron in certain applications where length of service was important.

## Damage to Notable Buildings

These original Horses of Saint Mark have been relocated indoors, and replaced by replicas.

In a 1987 article in *New Scientist,* Jack Harris reported that oxide jacking has caused significant damage to many historic structures in the United Kingdom, including St Paul's Cathedral, the British Museum and the Albert Memorial in London, Gloucester Cathedral, St. Margaret's Church in King's Lynn, Winchester Cathedral, and Blackburn Cathedral.

Harris also wrote that oxide jacking also damaged the ancient Horses of Saint Mark on the exterior of St. Mark's Basilica in Venice. Expansive rusting of iron and steel bolts and reinforcements affected the structural integrity of the copper horse sculptures, which were relocated indoors and replaced with replicas. Poorly-designed early 20th-century renovations also led to oxide jacking damage to the Acropolis of Athens.

Oxide jacking damage was discovered after a flood at the Farnsworth House.

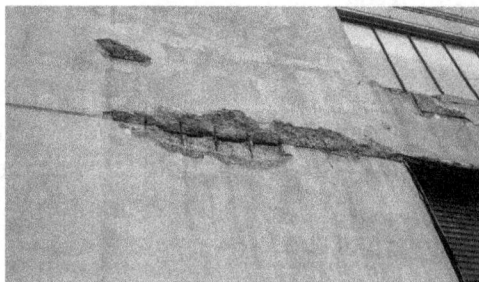

Oxide jacking has caused concrete spalling on walls
of the Herbst Pavilion at Fort Mason Center in San Franciscot

In the United States, rusting of iron pegs inserted into holes in the stone entrance stair in order to support handrails resulted in cracking of the steps at the Basilica of the Sacred Heart in Notre Dame, Indiana.

Oxide jacking damaged the terra cotta cornice on the Land Title Building in Philadelphia, designed in 1897 and expanded in 1902 by pioneer skyscraper architect Daniel Burnham. The Land Title complex, with its two interconnected towers, is on the National Register of Historic Places. By 1922, experts on architectural terra cotta were warning that the rusting of embedded iron fasteners could cause decorative building components to fail. This 1902 cornice is nearly 9 feet (2.7 m) high, projects 7 feet (2.1 m) from the facade of the building and is 465 feet (142 m) long. The cornice was stabilized, steel anchors subject to rusting were replaced with new stainless steel anchors, and the cornice was completely renovated. The project was completed in 1991.

Flooding in 2007 damaged the modernist Farnsworth House in Plano, Illinois, designed in 1945 by Ludwig Mies van der Rohe, and now owned by the National Trust for Historic Preservation. Among the damage discovered by an architect inspecting the house in 2007 was oxide jacking at the corners of the house's steel framework. The house flooded again in 2008.

## Damage to Reinforced Concrete Bridges and Buildings

Structures built of concrete and reinforced with metal rebar are also subject to damage by oxide jacking. Expansion of corroded rebar causes spalling of the concrete. Structures exposed to a marine environment, or where salt is used for de-icing purposes, are especially susceptible to this type of damage.

Research in the 1960s showed that 22% of concrete bridge decks in Pennsylvania showed signs of spalling due to oxide jacking within four years of construction. Oxide jacking caused widespread damage to concrete council houses built in the United Kingdom in the post World War II era.

According to an expert in the field, the problem resulted in "intensive worldwide research into the causes and repair of reinforcement corrosion, which in turn led to a vast output of research papers, conferences and publications on the subject."

## Damage to Stone Countertops

Countertop components fabricated out of granite and other natural stones are sometimes reinforced with metal rods inserted into grooves cut into the underside of the stone, and bonded in place with various resins. This procedure is called "rodding" by countertop fabricators. Most commonly, these rods will be placed near sink cutouts to prevent cracking of the brittle stone countertop during transportation and installation. Data published by the Marble Institute of America shows that this technique results in a 600% increase in the deflection strength of the component.

However, if a metal rod subject to oxidation or other forms of corrosion is used, and moisture from a sink or faucet reaches the rod, oxide jacking can crack the countertop directly above the rod. Mild steel and some grades of aluminum rods are known to cause oxide jacking failures in granite countertops. Skilled stone repair professionals can disassemble the cracked stone, remove the metal rod, and reassemble the stone using various resins tinted to match the colors of the stone. This type of problem can be prevented by using reinforcing rods made of stainless steel or fiberglass in the rodding procedure.

# Corrosive Substance

The international pictogram for corrosive chemicals.

DOT corrosive label

A corrosive substance is one that will destroy and damage other substances with which it comes into contact. It may attack a great variety of materials, including metals and various organic compounds, but people are mostly concerned with its effects on living tissue: it causes chemical burns on contact and can lead to complications when ingested.

## Chemical Terms

The word corrosive is derived from the Latin verb *corrodere*, which means *to gnaw*, indicating how these substances seem to "gnaw" their way through flesh or other material. Sometimes the word *caustic* is used as a synonym but *caustic* generally refers only to strong bases, particularly alkalis, and not to acids, oxidizers, or other non-alkaline corrosives.

A low concentration of a corrosive substance is usually an irritant. Corrosion of non-living surfaces such as metals is a distinct process. For example, a water/air electrochemical cell corrodes iron to rust. In the Globally Harmonized System, both rapid corrosion of metals and chemical corrosion of skin qualify for the corrosive symbol.

Corrosives are different from poisons in that corrosives are immediately dangerous to the tissues they contact, whereas poisons may have systemic toxic effects that require time to become evident. Colloquially, corrosives may be called *poisons* but the concepts are technically distinct. However, there is nothing which precludes a corrosive from being a poison; there are substances that are both corrosives and poisons.

## Effects on Living Tissue

Common corrosives are either strong acids, strong bases, or concentrated solutions of certain weak acids or weak bases. They can exist as any state of matter, including liquids, solids, gases, mists or vapors.

Their action on living tissue (e.g. skin, flesh and cornea) is mainly based on acid-base reactions of amide hydrolysis, ester hydrolysis and protein denaturation. Proteins (chemically composed of amide bonds) are destroyed via amide hydrolysis while lipids (many of which have ester bonds) are decomposed by ester hydrolysis. These reactions lead to chemical burns and are the mechanism of the destruction posed by corrosives.

Some corrosives possess other chemical properties which may extend their corrosive effects on living tissue. For example, sulfuric acid ($H_2SO_4$) at a high concentration is also a strong dehydrating agent, capable of dehydrating carbohydrates and liberating extra heat. This results in secondary thermal burns in addition to the chemical burns and may speed up its decomposing reactions on the contact surface. Some corrosives, such as nitric acid and concentrated sulfuric acid, are strong oxidizing agents as well, which significantly contributes to the extra damage caused. Hydrofluoric acid does not necessarily cause noticeable damage upon contact, but produces tissue damage and toxicity after being painlessly absorbed. Zinc chloride solutions are capable of destroying cellulose and corroding through paper and silk since the zinc cations in the solutions specifically attack hydroxyl groups, acting as a Lewis acid. This effect is not restricted to acids; so strong a base as calcium oxide, which has a strong affinity for water (forming calcium hydroxide, itself a strong and corrosive base), also releases heat capable of contributing thermal burns as well as delivering the corrosive effects of a strong alkali to moist flesh.

In addition, some corrosive chemicals, mostly acids such as hydrochloric acid and nitric acid, are volatile and can emit corrosive mists upon contact with air. Inhalation can damage the respiratory tract.

Corrosive substances are most hazardous to eyesight. A drop of a corrosive may cause blindness within 2–10 seconds through opacification or direct destruction of the cornea.

Ingestion of corrosives can induce severe consequences, including serious damage of the gastrointestinal tract, which can lead to vomiting, severe stomach aches, and death.

## Common Types

Common corrosive chemicals are classified into:

- Acids

  - Strong acids – the most common are sulfuric acid, nitric acid and hydrochloric acid ($H_2SO_4$, $HNO_3$ and HCl, respectively).

  - Some concentrated weak acids, for example formic acid and acetic acid

  - Strong Lewis acids such as anhydrous aluminum chloride and boron trifluoride

  - Lewis acids with specific reactivity; e.g., solutions of zinc chloride

  - Extremely strong acids (superacids)

- Bases

  - Caustics or alkalis, such as sodium hydroxide, potassium hydroxide, and calcium hydroxide

  - Alkali metals in the metallic form (e.g. elemental sodium), and hydrides of alkali and alkaline earth metals, such as sodium hydride, function as strong bases and hydrate to give caustics

  - Extremely strong bases (superbases) such as alkoxides, metal amides (e.g. sodium amide) and organometallic bases such as butyllithium

  - Some concentrated weak bases, such as ammonia when anhydrous or in a concentrated solution

- Dehydrating agents such as concentrated sulfuric acid, phosphorus pentoxide, calcium oxide, anhydrous zinc chloride, also elemental alkali metals

- Strong oxidizers such as concentrated hydrogen peroxide

- Electrophilic halogens: elemental fluorine, chlorine, bromine and iodine, and electrophilic salts such as sodium hypochlorite or N-chloro compounds such as chloramine-T; halide ions are not corrosive, except for fluoride

- Organic halides and organic acid halides such as acetyl chloride and benzyl chloroformate

- Acid anhydrides

- Alkylating agents such as dimethyl sulfate

- Some organic materials such as phenol ("carbolic acid")

## Personal Protective Equipment

Use of personal protective equipment, including items such as protective gloves, protective aprons, acid suits, safety goggles, a face shield, or safety shoes, is normally recommended when handling corrosive substances. Users should consult a safety data sheet for the specific recommendation for the corrosive substance of interest. The material of construction of the personal protective equipment is of critical importance as well. For example, although rubber gloves and rubber aprons may be made out of a chemically resistant elastomer such as nitrile rubber, neoprene, or butyl rubber, each of these materials has different resistance to different corrosives and they should not be substituted for each other.

## Uses

Some corrosive chemicals are valued for various uses, the most common of which is in household cleaning agents. For example, most drain cleaners contain either acids or alkalis due to their capabilities of dissolving greases and proteins inside water pipes such as limescale.

In chemical uses, high chemical reactivity is often desirable, as the rates of chemical reactions depend on the activity (effective concentration) of the reactive species. For instance, catalytic sulfuric acid is used in the alkylation process in an oil refinery: the activity of carbocations, the reactive intermediate, is higher with stronger acidity, and thus the reaction proceeds faster. Once used, corrosives are most often recycled or neutralized. However, there have been environmental problems with untreated corrosive effluents or accidental discharges.

## References

- Bucher, Ward (1996). Dictionary of Building Preservation. New York: Wiley Interscience. p. 319. ISBN 0-471-14413-4

- Y.Yan, Biotribocorrosion – an appraisal of the time dependence of wear and corrosion interactions Part II: Surface analysis. Journal of Physics D: Applied Physics. 39(2006) p.3206-3212

- Godfrey, Douglas (2003). "Fretting corrosion or false brinelling?" (PDF). ribology and Lubrication Technology. 59 (12): 28–31. Retrieved 23.06.2017

- Munnings, C.; Badwal, S. P. S.; Fini, D. (20 February 2014). "Spontaneous stress-induced oxidation of Ce ions in Gd-doped ceria at room temperature". Ionics. 20 (8): 1117–1126. doi:10.1007/s11581-014-1079-2

- Winkler, Erhard M. (1997). Stone in Architecture: Properties, Durability (3rd ed.). Berlin: Springer-Verlag. p. 239. ISBN 3-540-57626-6

- Errichello, Robert (2004). "Another perspective: false brinelling and fretting corrosion". Tribology & lubrication technology. 60 (4): 34-36. Retrieved 23.06.2017

- Pullar-Strecker, Peter (2002). Concrete reinforcement corrosion: from assessment to repair decisions. London: Thomas Telford Publishing. ISBN 978-0-7277-3182-1

- Schwack, Fabian (2016). "Comparison of Life Calculations for Oscillating Bearings Considering Individual Pitch Control in Wind Turbines". Journal of Physics: Conference Series. 753 (11): 1–10. doi:10.1088/1742-6596/753/11/112013. Retrieved 23.03.2016

- Hill, C.W. (October 1922). "Terra Cotta Problems Suggested for Discussion and Investigation". Journal of the American Ceramic Society. Easton, Pennsylvania: American Ceramic Society. 5 (10): 732–38. doi:10.1111/j.1151-2916.1922.tb17607.x. Retrieved January 2, 2012

# Preventive Measures and Measurements of Corrosion

Conversion coatings are the coatings given to metals that protect them from corrosion. The other ways of preventing corrosion are corrosion inhibitor, cathodic protection, galvanization, coating, low plasticity burnishing, corrosion mapping by ultrasonics, cyclic corrosion testing and salt spray test. The topics discussed in the chapter are of great importance to broaden the existing knowledge on corrosion engineering.

## Anti-corrosion

Anti-corrosion refers to the protection of metal surfaces from corroding in high-risk (corrosive) environments.

When metallic materials are put into corrosive environments, they tend to have chemical reactions with the air and/or water. The effects of corrosion become evident on the surfaces of these materials. For example, after putting a piece of iron into a corrosive atmosphere for an extended period, it starts rusting due to oxygen interaction with water on the surface of the piece of iron.

Metal equipment lacking any preventive (anti-corrosive) measures may become rusted both inside and out, depending upon atmospheric conditions and how much of that equipment is exposed to the air. There are a number of methods for preventing corrosion, especially in marine applications. Anti-corrosion measures are of particular importance in environments where high humidity, mist, and salt are factors.

### Protection from Corrosion

- Anti-corrosion materials

  o Stainless steel

  o Electrogalvanized cold-rolled steel (SECC), a type of zinc-plated steel, used as a housing to protect electronic components

  o Barrier coatings provide a consistent film that forms a chemically resistant barrier and blocks external factors from causing harm.

- o Sacrificial coatings contain certain element (such as aluminum or zinc) which oxidize sacrificially to ensure the protected element remains corrosion free.

- Enclosure design: fully sealed enclosures (at least full IP65) to protect internal electronic components, used in highly salty and damp places

- Painting treatments on the surface: Another way to protect metal housings from corrosion is by using anti-corrosive paint or powder coat on the metallic surface. The function of this coating is to act as a barrier that inhibits contact between chemical compounds or corrosive materials with the metal housing.

## Stainless Steel

Stainless steel cladding is used on the Walt Disney Concert Hall

In metallurgy, stainless steel, also known as inox steel or inox from French *inoxydable* (inoxidizable), is a steel alloy with a minimum of 10.5% chromium content by mass.

Stainless steel is notable for its corrosion resistance, and it is widely used for food handling and cutlery among many other applications.

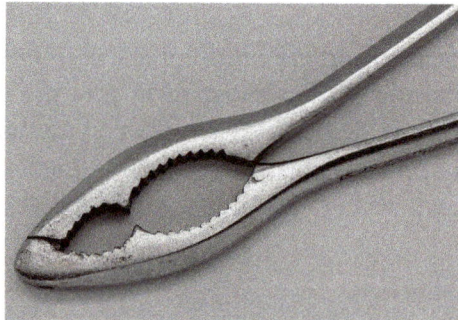

Stainless steel is used for corrosion-resistant tools such as this nutcracker

Stainless steel does not readily corrode, rust or stain with water as ordinary steel does. However, it is not fully stain-proof in low-oxygen, high-salinity, or poor air-circulation environments.

To reduce staining, the surface of stainless steel must be kept clean throughout its aging process when oxygen reacts with the surface to form a protective chromium-oxide layer. Once this process has taken place, the surface becomes much more resistant to staining.

There are various grades and surface finishes of stainless steel to suit the environment the alloy must endure. Stainless steel is used where both the properties of steel and corrosion resistance are required.

Stainless steel differs from carbon steel by the amount of chromium present. Unprotected carbon steel rusts readily when exposed to air and moisture. This iron oxide film (the rust) is active and accelerates corrosion by making it easier for more iron oxide to form. Since iron oxide has lower density than steel, the film expands and tends to flake and fall away. In comparison, stainless steels contain sufficient chromium to undergo passivation, forming an inert film of chromium oxide on the surface. This layer prevents further corrosion by blocking oxygen diffusion to the steel surface and stops corrosion from spreading into the bulk of the metal. Passivation occurs only if the proportion of chromium is high enough and oxygen is present in it.

Stainless steel's resistance to corrosion and staining, low maintenance, and familiar lustre make it an ideal material for many applications. The alloy is milled into coils, sheets, plates, bars, wire, and tubing to be used in cookware, cutlery, household hardware, surgical instruments, major appliances, industrial equipment (for example, in sugar refineries) and as an automotive and aerospace structural alloy and construction material in large buildings. Storage tanks and tankers used to transport orange juice and other food are often made of stainless steel, because of its corrosion resistance. This also influences its use in commercial kitchens and food processing plants, as it can be steam-cleaned and sterilized and does not need paint or other surface finishes.

## Properties

Stainless steel (row 3) resists salt-water corrosion better than aluminum-bronze (row 1) or copper-nickel alloys (row 2)

## Oxidation

High oxidation resistance in air at ambient temperature is normally achieved with addition of a minimum of 13% (by weight) chromium, and up to 26% is used for harsh environments. The chromium forms a passivation layer of chromium(III) oxide ($Cr_2O_3$) when exposed to oxygen. The layer is too thin to be visible, and the metal remains lustrous and smooth. The layer is impervious to water and air, protecting the metal beneath, and this layer quickly reforms when the surface is scratched. This phenomenon is called passivation and is seen in other metals, such as aluminium and titanium. Corrosion resistance can be adversely affected if the component is used in a non-oxygenated environment, a typical example being underwater keel bolts buried in timber.

When stainless steel parts such as nuts and bolts are forced together, the oxide layer can be scraped off, allowing the parts to weld together. When forcibly disassembled, the welded material may be torn and pitted, a destructive effect known as galling. Galling can be avoided by the use of dissimilar materials for the parts forced together, for example bronze and stainless steel, or even different types of stainless steels (martensitic against austenitic). However, two different alloys electrically connected in a humid, even mildly acidic environment may act as a voltaic pile and corrode faster. Nitronic alloys, made by selective alloying with manganese and nitrogen, may have a reduced tendency to gall. Additionally, threaded joints may be lubricated to provide a film between the two parts and prevent galling. Low-temperature carburizing is another option that virtually eliminates galling and allows the use of similar materials without the risk of corrosion and the need for lubrication.

## Acids

Stainless steel is generally highly resistant to attack from acids, but this quality depends on the kind and concentration of the acid, the surrounding temperature, and the type of steel. Type 904 is resistant to sulfuric acid at room temperature, even in high concentrations; types 316 and 317 are resistant below 10%; and type 304 should not be used in the presence of sulfuric acid at any concentration. All types of stainless steel resist attack from phosphoric acid, types 316 and 317 more so than 304; types 304L and 430 have been successfully used with nitric acid. Hydrochloric acid will damage any kind of stainless steel, and should be avoided.

## Bases

The 300 series of stainless steel grades is unaffected by any of the weak bases such as ammonium hydroxide, even in high concentrations and at high temperatures. The same grades of stainless exposed to stronger bases such as sodium hydroxide at high concentrations and high temperatures will likely experience some etching and cracking, especially with solutions containing chlorides such as sodium hypochlorite.

## Organics

Types 316 and 317 are both useful for storing and handling acetic acid, especially in solutions where it is combined with formic acid and when aeration is not present (oxygen helps protect stainless steel under such conditions), though 317 provides the greatest level of resistance to corrosion. Type 304 is also commonly used with formic acid though it will tend to discolor the solution. All grades resist damage from aldehydes and amines, though in the latter case grade 316 is preferable to 304; cellulose acetate will damage 304 unless the temperature is kept low. Fats and fatty acids only affect grade 304 at temperatures above 150 °C (302 °F), and grade 316 above 260 °C (500 °F), while 317 is unaffected at all temperatures. Type 316L is required for processing of urea.

## Electricity and Magnetism

left nut is not in inox and is rusty

Poor selection of materials can cause galvanic corrosion to other metals in contact with stainless steel

Like steel, stainless steel is a relatively poor conductor of electricity, with significantly lower electrical conductivity than copper. Other metals in contact with stainless steel, particularly in a damp or acidic environment, may suffer galvanic corrosion even though the stainless metal may be unaffected.

Ferritic and martensitic stainless steels are magnetic. Annealed austenitic stainless steels are non-magnetic. Work hardening can make austenitic stainless steels slightly magnetic.

## History

The corrosion resistance of iron-chromium alloys was first recognized in 1821 by French metallurgist Pierre Berthier, who noted their resistance against attack by some acids and suggested their use in cutlery. Metallurgists of the 19th century were unable to produce the combination of low carbon and high chromium found in most modern stainless steels, and the high-chromium alloys they could produce were too brittle to be practical.

### A NON-RUSTING STEEL.

**Sheffield Invention Especially Good for Table Cutlery.**

According to Consul John M. Savage, who is stationed at Sheffield, England, a firm in that city has introduced a stainless steel, which is claimed to be non-rusting, unstainable, and untarnishable. This steel is said to be especially adaptable for table cutlery, as the original polish is maintained after use, even when brought in contact with the most acid foods, and it requires only ordinary washing to cleanse.

"It is claimed," writes Mr. Savage in the Commerce Reports, "that this steel retains a keen edge much like that of the best double-sheer steel, and, as the properties claimed are inherent in the steel and not due to any treatment, knives can readily be sharpened on a 'steel' or by using the ordinary cleaning machine or knifeboard. It is expected it will prove a great boon, especially to large users of cutlery, such as hotels, steamships, and restaurants.

"The price of this steel is about 26 cents a pound for ordinary sizes, which is about double the price of the usual steel for the same purpose. It also costs more to work up, so that the initial cost of articles made from this new discovery, it is estimated, will be about double the present cost; but it is considered that the saving of labor to the customer will more than cover the total cost of the cutlery in the first twelve month."

An announcement, as it appeared in the 1915 *New York Times*, of the development of stainless steel in Sheffield, England.

In 1872, the Englishmen Clark and Woods patented an alloy that would today be considered a stainless steel.

In the late 1890s Hans Goldschmidt of Germany developed an aluminothermic (thermite) process for producing carbon-free chromium. Between 1904 and 1911 several researchers, particularly Leon Guillet of France, prepared alloys that would today be considered stainless steel.

Friedrich Krupp Germaniawerft built the 366-ton sailing yacht *Germania* featuring a chrome-nickel steel hull in Germany in 1908. In 1911, Philip Monnartz reported on the relationship between chromium content and corrosion resistance. On 17 October 1912, Krupp engineers Benno Strauss and Eduard Maurer patented austenitic stainless steel as Nirosta.

Similar developments were taking place contemporaneously in the United States, where Christian Dantsizen and Frederick Becket were industrializing ferritic stainless steel. In 1912, Elwood Haynes applied for a US patent on a martensitic stainless steel alloy, which was not granted until 1919.

Monument to Harry Brearley at the former Brown Firth Research Laboratory in Sheffield, England

In 1912, Harry Brearley of the Brown-Firth research laboratory in Sheffield, England, while seeking a corrosion-resistant alloy for gun barrels, discovered and subsequently industrialized a martensitic stainless steel alloy. The discovery was announced two years later in a January 1915 newspaper article in *The New York Times*. The metal was later marketed under the "Staybrite" brand by Firth Vickers in England and was used for the new entrance canopy for the Savoy Hotel in London in 1929. Brearley applied for a US patent during 1915 only to find that Haynes had already registered a patent. Brearley and Haynes pooled their funding and with a group of investors formed the American Stainless Steel Corporation, with headquarters in Pittsburgh, Pennsylvania.

In the beginning stainless steel was sold in the US under different brand names like "Allegheny metal" and "Nirosta steel". Even within the metallurgy industry the eventual name remained unsettled; in 1921 one trade journal was calling it "unstainable steel". In 1929, before the Great Depression hit, over 25,000 tons of stainless steel were manufactured and sold in the US.

## Types

Pipes and fittings made of stainless steel

There are different types of stainless steels: when nickel is added, for instance, the austenite structure of iron is stabilized. This crystal structure makes such steels virtually non-magnetic and less brittle at low temperatures. For greater hardness and strength, more carbon is added. With proper heat treatment, these steels are used for such products as razor blades, cutlery, and tools.

Significant quantities of manganese have been used in many stainless steel compositions. Manganese preserves an austenitic structure in the steel, similar to nickel, but at a lower cost.

Stainless steels are also classified by their crystalline structure:

- *Austenitic*, or 200 and 300 series, stainless steels have an austenitic crystalline structure, which is a face-centered cubic crystal structure. Austenite steels make up over 70% of total stainless steel production. They contain a maximum of 0.15% carbon, a minimum of 16% chromium, and sufficient nickel and/or manganese to retain an austenitic structure at all temperatures from the cryogenic region to the melting point of the alloy.

- 200 Series—austenitic chromium-nickel-manganese alloys. Type 201 is hardenable through cold working; Type 202 is a general purpose stainless steel. Decreasing nickel content and increasing manganese results in weak corrosion resistance.

- 300 Series. The most widely used austenite steel is the 304, also known as *18/8* for its composition of 18% chromium and 8% nickel. 304 may be referred to as A2 stainless. The second most common austenite steel is the 316 grade, also referred to as A4 stainless and called marine grade stainless, used primarily for its increased resistance to corrosion. A typical composition of 18% chromium and 10% nickel, commonly known as *18/10 stainless,* is often used in cutlery and high-quality cookware. *18/0* is also available.

*Superaustenitic* stainless steels, such as Allegheny Ludlum alloy AL-6XN and 254SMO, exhibit great resistance to chloride pitting and crevice corrosion because of high molybdenum content (>6%) and nitrogen additions, and the higher nickel content ensures better resistance to stress-corrosion cracking versus the 300 series. The higher alloy content of superaustenitic steels makes them more expensive. Other steels can offer similar performance at lower cost and are preferred in certain applications. For example ASTM A387 is used in pressure vessels but is a low-alloy carbon steel with a chromium content of 0.5% to 9%. Low-carbon versions, for example 316L or 304L, are used to avoid corrosion problems caused by welding. Grade 316LVM is preferred where biocompatibility is required (such as body implants and piercings). The "L" means that the carbon content of the alloy is below 0.03%, which reduces the sensitization effect (precipitation of chromium carbides at grain boundaries) caused by the high temperatures involved in welding.

- *Ferritic* stainless steels generally have better engineering properties than austenitic grades, but have reduced corrosion resistance, because of the lower chromium and nickel content. They are also usually less expensive. Ferritic stainless steels have a body-centered cubic crystal system and contain between 10.5% and 27% chromium with very little nickel, if any, but some types can contain lead. Most compositions include molybdenum; some, aluminium or titanium. Common ferritic grades include 18Cr-2Mo, 26Cr-1Mo, 29Cr-4Mo, and 29Cr-4Mo-2Ni. These alloys can be degraded by the presence of chromium, an intermetallic phase which can precipitate upon welding.

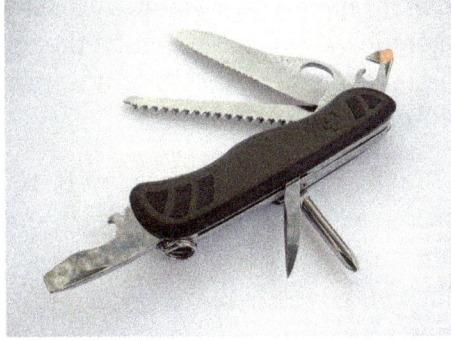

Swiss Army knives are made of martensitic stainless steel.

- *Martensitic* stainless steels are not as corrosion-resistant as the other two classes but are extremely strong and tough, as well as highly machinable, and can be hardened by heat treatment. Martensitic stainless steel contains chromium (12–14%), molybdenum (0.2–1%), nickel (less than 2%), and carbon (about 0.1–1%) (giving it more hardness but making the material a bit more brittle). It is quenched and magnetic.

- *Duplex steel* stainless steels have a mixed microstructure of austenite and ferrite, the aim usually being to produce a 50/50 mix, although in commercial alloys the ratio may be 40/60. Duplex stainless steels have roughly twice the strength compared to austenitic stainless steels and also improved resistance to localized corrosion, particularly pitting, crevice corrosion and stress corrosion cracking. They are characterized by high chromium (19–32%) and molybdenum (up to 5%) and lower nickel contents than austenitic stainless steels.

The properties of duplex stainless steels are achieved with an overall lower alloy content than similar-performing super-austenitic grades, making their use cost-effective for many applications. Duplex grades are characterized into groups based on their alloy content and corrosion resistance.

- *Lean duplex* refers to grades such as UNS S32101 (LDX 2101), S32202 (UR2202), S32304, and S32003.

- *Standard duplex* refers to grades with 22% chromium, such as UNS S31803/ S32205, with 2205 being the most widely used.

- *Super duplex* is by definition a duplex stainless steel with a Pitting Resistance Equivalent Number (PREN) > 40, where PREN = %Cr + 3.3x(%Mo + 0.5x%W) + 16x%N. Usually super duplex grades have 25% or more chromium. Some common examples are S32760 (Zeron 100 via Rolled Alloys), S32750 (2507), and S32550 (Ferralium 255 via Langley Alloys).

- *Hyper duplex* refers to duplex grades with a PRE > 48. UNS S32707 and S33207 are the only grades currently available on the market.

- *Precipitation-hardening martensitic* stainless steels have corrosion resistance comparable to austenitic varieties, but can be precipitation hardened to even higher strengths than the other martensitic grades. The most common, 17-4PH, uses about 17% chromium and 4% nickel.

The designation "CRES" is used in various industries to refer to corrosion-resistant steel. Most mentions of CRES refer to stainless steel, although the correspondence is not absolute, because there are other materials that are corrosion-resistant but not stainless steel.

## Grades

There are over 150 grades of stainless steel, of which 15 are most commonly used. There are a number of systems for grading stainless and other steels, including US SAE steel grades.

## Comparison of Standardized Steels

| EN-standard Steel no. k.h.s DIN | EN-standard Steel name | SAE grade | UNS |
|---|---|---|---|
| 1.4512 | X6CrTi12 | 409 | S40900 |
| | | 410 | S41000 |
| 1.4016 | X6Cr17 | 430 | S43000 |
| 1.4109 | X65CrMo14 | 440A | S44002 |
| 1.4112 | X90CrMoV18 | 440B | S44003 |
| 1.4125 | X105CrMo17 | 440C | S44004 |
| | | 440F | S44020 |
| 1.4310 | X10CrNi18-8 | 301 | S30100 |
| 1.4318 | X2CrNiN18-7 | 301LN | |
| 1.4301 | X5CrNi18-10 | 304 | S30400 |
| 1.4307 | X2CrNi18-9 | 304L | S30403 |
| 1.4306 | X2CrNi19-11 | 304L | S30403 |
| 1.4311 | X2CrNiN18-10 | 304LN | S30453 |
| 1.4948 | X6CrNi18-11 | 304H | S30409 |
| 1.4303 | X5CrNi18-12 | 305 | S30500 |
| | X5CrNi30-9 | 312 | |
| 1.4841 | X22CrNi2520 | 310 | S31000 |
| 1.4845 | X 5 CrNi 2520 | 310S | S31008 |
| 1.4401 | X5CrNiMo17-12-2 | 316 | S31600 |
| 1.4408 | G-X 6 CrNiMo 18-10 | 316 | S31600 |
| 1.4436 | X3CrNiMo17-13-3 | 316 | S31600 |
| 1.4406 | X2CrNiMoN17-12-2 | 316LN | S31653 |

| EN-standard Steel no. k.h.s DIN | EN-standard Steel name | SAE grade | UNS |
|---|---|---|---|
| 1.4404 | X2CrNiMo17-12-2 | 316L | S31603 |
| 1.4432 | X2CrNiMo17-12-3 | 316L | S31603 |
| 1.4435 | X2CrNiMo18-14-3 | 316L | S31603 |
| 1.4571 | X6CrNiMoTi17-12-2 | 316Ti | S31635 |
| 1.4429 | X2CrNiMoN17-13-3 | 316LN | S31653 |
| 1.4438 | X2CrNiMo18-15-4 | 317L | S31703 |
| 1.4541 | X6CrNiTi18-10 | 321 | S32100 |
| 1.4878 | X12CrNiTi18-9 | 321H | S32109 |
| 1.4362 | X2CrNi23-4 | 2304 | S32304 |
| 1.4462 | X2CrNiMoN22-5-3 | 2205 | S31803/S32205 |
| 1.4501 | X2CrNiMoCuWN25-7-4 | J405 | S32760 |
| 1.4539 | X1NiCrMoCu25-20-5 | 904L | N08904 |
| 1.4529 | X1NiCrMoCuN25-20-7 | | N08926 |
| 1.4547 | X1CrNiMoCuN20-18-7 | 254SMO | S31254 |

## Standard Finishes

316L stainless steel, with an unpolished, mill finish

Standard mill finishes can be applied to flat rolled stainless steel directly by the rollers and by mechanical abrasives. Steel is first rolled to size and thickness and then annealed to change the properties of the final material. Any oxidation that forms on the surface (mill scale) is removed by pickling, and a passivation layer is created on the surface. A final finish can then be applied to achieve the desired aesthetic appearance.

- No. 0: Hot rolled, annealed, thicker plates

- No. 1: Hot rolled, annealed and passivated

- No. 2D: Cold rolled, annealed, pickled and passivated

- No. 2B: Same as above with additional pass through highly polished rollers

- No. 2BA: Bright annealed (BA or 2R) same as above then bright annealed under oxygen-free atmospheric condition

- No. 3: Coarse abrasive finish applied mechanically

- No. 4: Brushed finish

- No. 5: Satin finish

- No. 6: Matte finish (brushed but smoother than #4)

- No. 7: Reflective finish

- No. 8: Mirror finish

- No. 9: Bead blast finish

- No. 10: Heat colored finish—offering a wide range of electropolished and heat colored surfaces

## Applications

The 630-foot-high (190 m), stainless-clad (type 304)
Gateway Arch defines St. Louis's skyline

The pinnacle of New York's Chrysler Building is clad
with *Nirosta* stainless steel, a form of Type 302

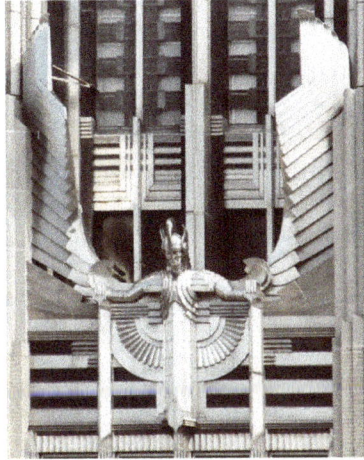

An art deco sculpture on the Niagara-Mohawk Power building in Syracuse, New York

Stainless steel is often used for cookware

## Architecture

Stainless steel is used for buildings for both practical and aesthetic reasons. Stainless steel was in vogue during the art deco period. The most famous example of this is the upper portion of the Chrysler Building (pictured). Some diners and fast-food restaurants use large ornamental panels and stainless fixtures and furniture. Because of the durability of the material, many of these buildings still retain their original appearance. Stainless steel is used today in building construction because of its durability and because it is a weldable building metal that can be made into aesthetically pleasing shapes. An example of a building in which these properties are exploited is the Art Gallery of Alberta in Edmonton, which is wrapped in stainless steel.

Type 316 stainless is used on the exterior of both the Petronas Twin Towers and the Jin Mao Building, two of the world's tallest skyscrapers.

The Parliament House of Australia in Canberra has a stainless steel flagpole weighing over 220 tonnes (240 short tons).

The aeration building in the Edmonton Composting Facility, the size of 14 hockey rinks, is the largest stainless steel building in North America.

## Bridges

The Helix Bridge is a pedestrian bridge linking Marina Centre with Marina South in the Marina Bay area in Singapore.

- Cala Galdana Bridge in Minorca (Spain) was the first stainless steel road bridge.

- Sant Fruitos Pedestrian Bridge (Catalonia, Spain), arch pedestrian bridge.

- Padre Arrupe Bridge (Bilbao, Spain) links the Guggenheim museum to the University of Deusto.

## Monuments and sculptures

- Unisphere, constructed as the theme symbol of the 1964 New York World's Fair, is constructed of Type 304L stainless steel as a spherical framework with a diameter of 120 feet (37 m) (New York City)

- Gateway Arch (pictured) is clad entirely in stainless steel: 886 tons (804 metric tonnes) of 0.25 in (6.4 mm) plate, #3 finish, type 304 stainless steel. (St. Louis, Missouri)

- United States Air Force Memorial has an austenitic stainless steel structural skin (Arlington, Virginia)

- Atomium was renovated with stainless-steel cladding in a renovation completed in 2006; previously the spheres and tubes of the structure were clad in aluminium (Brussels, Belgium)

- Cloud Gate sculpture by Anish Kapoor (Chicago, Illinois)

- Sibelius Monument is made entirely of stainless steel tubes (Helsinki, Finland)

- The Kelpies (Falkirk, Scotland)

- Man of Steel (sculpture) under construction (Rotherham, England)

- Juraj Jánošík monument (Terchova, Slovakia)

## Airports

Stainless steel is a modern trend for roofing material for airports due to its low glare reflectance to keep pilots from being blinded, also for its properties that allow thermal reflectance in order to keep the surface of the roof close to ambient temperature. The Hamad International Airport in Qatar was built with all stainless steel roofing for these reasons, as well as the Sacramento International Airport in California.

# Locomotion

## Automotive bodies

The Allegheny Ludlum Corporation worked with Ford on various concept cars with stainless steel bodies from the 1930s through the 1970s to demonstrate the material's potential. The 1957 and 1958 Cadillac Eldorado Brougham had a stainless steel roof. In 1981 and 1982, the DeLorean DMC-12 production automobile used Type-304 stainless steel body panels over a glass-reinforced plastic monocoque. Intercity buses made by Motor Coach Industries are partially made of stainless steel. The aft body panel of the Porsche Cayman model (2-door coupe hatchback) is made of stainless steel. It was discovered during early body prototyping that conventional steel could not be formed without cracking (due to the many curves and angles in that automobile). Thus, Porsche was forced to use stainless steel on the Cayman.

Some automotive manufacturers use stainless steel as decorative highlights in their vehicles.

## Passenger rail cars

Rail cars have commonly been manufactured using corrugated stainless steel panels (for additional structural strength). This was particularly popular during the 1960s and 1970s, but has since declined. One notable example was the early Pioneer Zephyr. Notable former manufacturers of stainless steel rolling stock included the Budd Company (USA), which has been licensed to Japan's Tokyu Car Corporation, and the Portuguese company Sorefame. Many railcars in the United States are still manufactured with stainless steel, unlike other countries who have shifted away.

## Aircraft

Budd also built two airplanes, the Budd BB-1 Pioneer and the Budd RB-1 Conestoga, of stainless steel tube and sheet. The first, which had fabric wing coverings, is on display at the Franklin Institute, being the longest continuous display of an aircraft ever, since 1934. The RB-2 Was almost all stainless steel, save for the control surfaces. One survives at the Pima Air & Space Museum, adjacent to Davis–Monthan Air Force Base.

The American Fleetwings Sea Bird amphibious aircraft of 1936 was also built using a spot-welded stainless steel hull.

Due to its thermal stability, the Bristol Aeroplane Company built the all-stainless steel Bristol 188 high-speed research aircraft, which first flew in 1963. However, the practical problems encountered meant that Concorde employed aluminium alloys.

The use of stainless steel in mainstream aircraft is hindered by its excessive weight compared to other materials, such as aluminium.

## Medicine

Surgical tools and medical equipment are usually made of stainless steel, because of its durability and ability to be sterilized in an autoclave. In addition, surgical implants such as bone reinforcements and replacements (e.g. hip sockets and cranial plates) are made with special alloys formulated to resist corrosion, mechanical wear, and biological reactions *in vivo*.

Stainless steel is used in a variety of applications in dentistry. It is common to use stainless steel in many instruments that need to be sterilized, such as needles, endodontic files in root canal therapy, metal posts in root canal–treated teeth, temporary crowns and crowns for deciduous teeth, and arch wires and brackets in orthodontics. The surgical stainless steel alloys (e.g., 316 low-carbon steel) have also been used in some of the early dental implants.

## Culinary use

Stainless steel is often preferred for kitchen sinks because of its ruggedness, durability, heat resistance, and ease of cleaning. In better models, acoustic noise is controlled by applying resilient undercoating to dampen vibrations. The material is also used for cladding of surfaces such as appliances and backsplashes.

Cookware and bakeware may be clad in stainless steels, to enhance their cleanability and durability, and to permit their use in induction cooking. Because stainless steel is a poor conductor of heat, it is often used as a thin surface cladding over a core of copper or aluminum, which conduct heat more readily.

Cutlery is normally stainless steel, both for low corrosion, ease of cleaning, negligible toxicity, as well as not flavoring the food by electrolytic activity.

## Jewelry

Stainless steel is used for jewelry and watches, with 316L being the type commonly used for such applications. It can be re-finished by any jeweler and will not oxidize or turn black.

Valadium, a stainless steel and 12% nickel alloy is used to make class and military rings. Valadium is usually silver-toned, but can be electro-plated to give it a gold tone. The gold tone variety is known as Sun-lite Valadium. Other "Valadium" types of alloy are trade-named differently, with such names as "Siladium" and "White Lazon".

## Firearms

Some firearms incorporate stainless steel components as an alternative to blued or parkerized steel. Some handgun models, such as the Smith & Wesson Model 60 and the Colt M1911 pistol, can be made entirely from stainless steel. This gives a high-luster finish similar in appearance to nickel plating. Unlike plating, the finish is not subject to flaking, peeling, wear-off from rubbing (as when repeatedly removed from a holster), or rust when scratched.

## 3D Printing

Some 3D printing providers have developed proprietary stainless steel sintering blends for use in rapid prototyping. One of the more popular stainless steel grades used in 3D printing is 316L stainless steel. Due to the high temperature gradient and fast rate of solidification, stainless steel products manufactured via 3D printing tend to have a more refined microstructure; this in turn results in better mechanical properties. However, stainless steel is not used as much as materials like Ti6Al4V in the 3D printing industry; this is because manufacturing stainless steel products via traditional methods is currently much more economically competitive.

## Recycling and Reusing

Stainless steel is 100% recyclable. An average stainless steel object is composed of about 60% recycled material of which approximately 40% originates from end-of-life products and about 60% comes from manufacturing processes. According to the International Resource Panel's Metal Stocks in Society report, the per capita stock of stainless steel in use in society is 80–180 kg in more developed countries and 15 kg in less-developed countries.

There is a secondary market that recycles usable scrap for many stainless steel markets. The product is mostly coil, sheet, and blanks. This material is purchased at a less-than-prime price and sold to commercial quality stampers and sheet metal houses. The material may have scratches, pits, and dents but is made to the current specifications.

## Nanoscale Stainless Steel

Stainless steel nanoparticles have been produced in lab settings. This synthesis uses oxidative Kirkendall Diffusion to build a thin protective barrier which prevent further oxidation. These may have applications as additives for high performance applications.

## Health Effects

Stainless steel is generally considered to be biologically inert, but some sensitive individuals develop a skin irritation due to a nickel allergy caused by certain alloys.

# Conversion Coating

Conversion coatings are coatings for metals where the part surface is converted into the coating with a chemical or electro-chemical process. Examples include chromate conversion coatings, phosphate conversion coatings, bluing, black oxide coatings on steel, and anodizing. They are used for corrosion protection, to add decorative color and as paint primers.

## Phosphate Conversion Coating

Phosphate coatings are used on steel parts for corrosion resistance, lubricity, or as a foundation for subsequent coatings or painting. It serves as a conversion coating in which a dilute solution of phosphoric acid and phosphate salts is applied via spraying or immersion and chemically reacts with the surface of the part being coated to form a layer of insoluble, crystalline phosphates. Phosphate conversion coatings can also be used on aluminium, zinc, cadmium, silver and tin.

The main types of phosphate coatings are manganese, iron and zinc. Manganese phosphates are used both for corrosion resistance and lubricity and are applied only by immersion. Iron phosphates are typically used as a base for further coatings or painting and are applied by immersion or by spraying. Zinc phosphates are used for corrosion resistance (phosphate and oil), a lubricant base layer, and as a paint/coating base and can also be applied by immersion or spraying.

## Process

The application of phosphate coatings makes use of phosphoric acid and takes advantage of the low solubility of phosphates in medium or high pH solutions. Iron, zinc or manganese phosphate salts are dissolved in a solution of phosphoric acid. When steel or iron parts are placed in the phosphoric acid, a classic acid and metal reaction takes place which locally depletes the hydronium ($H_3O^+$) ions, raising the pH, and causing the dissolved salt to fall out of solution and be precipitated on the surface. The acid and metal reaction also creates iron phosphate locally which may also be deposited. In the case of depositing zinc phosphate or manganese phosphate the additional iron phosphate is frequently an undesirable addition to the coating.

The acid and metal reaction also generates hydrogen gas in the form of tiny bubbles that adhere to the surface of the metal. These prevent the acid from reaching the metal surface and slows down the reaction. To overcome this sodium nitrite is frequently added to act as an oxidizing agent that reacts with the hydrogen to form water. This chemistry is known as a nitrate accelerated solution. Hydrogen is prevented from forming a passivating layer on the surface by the oxidant additive.

The following is a typical phosphating procedure:

1. cleaning the surface

2. rinsing

3. surface activation

4. phosphating

5. rinsing

6. neutralizing rinse (optional)

7. drying

8. application of supplemental coatings: lubricants, sealers, oil, etc.

The performance of the phosphate coating is significantly dependent on the crystal structure as well as the weight. For example, a microcrystalline structure is usually optimal for corrosion resistance or subsequent painting. A coarse grain structure impregnated with oil, however, may be the most desirable for wear resistance. These factors are controlled by selecting the appropriate phosphate solution, using various additives, and controlling bath temperature, concentration, and phosphating time. A widely used additive is to seed the metal surface with tiny particles of titanium salts by adding these to the rinse bath preceding the phosphating. This is known as activation.

## Uses

Phosphate coatings are often used to provide corrosion resistance, however, phosphate coatings on their own do not provide this because the coating is porous. Therefore, oil or other sealers are used to achieve corrosion resistance. Zinc and manganese coatings are used to help break in components subject to wear and help prevent galling.

Most phosphate coatings serve as a surface preparation for further coating and/or painting, a function it performs effectively with excellent adhesion and electric isolation. The porosity allows the additional materials to seep into the phosphate coating and become mechanically interlocked after drying. The dielectric nature will electrically isolate anodic and cathodic areas on the surface of the part, minimizing underfilm corrosion that sometimes occurs at the interface of the paint/coating and the substrate.

Zinc phosphate coatings are frequently used in conjunction with sodium stearate (soap) to form a lubrication layer in cold and hot forging. The sodium stearate reacts with the phosphate crystal which in turn are strongly bonded to the metal surface. The reacted soap layer then forms a base for additional unreacted soap to be deposited on top so that a thick three part coating of zinc phosphate, reacted soap and unreacted soap is built up. The resulting coating remains adhered to the metal surface even under extreme deformation. The zinc phosphate is in fact abrasive and it is the soap which performs the actual lubrication. The soap layer must be thick enough to prevent substantial contact between the metal forming dies and phosphate crystal.

## Black Oxide

Black oxide or blackening is a conversion coating for ferrous materials, stainless steel, copper and copper based alloys, zinc, powdered metals, and silver solder. It is used to add mild corrosion resistance, for appearance and to minimize light reflection. To achieve maximal corrosion resistance the black oxide must be impregnated with oil or wax. One of its advantages over other coatings is its minimal buildup.

## Ferrous Materials

### Hot Black Oxide

Hot baths of sodium hydroxide, nitrates, and nitrites at 141 °C (286 °F) are used to convert the surface of the material into magnetite ($Fe_3O_4$). Water must be periodically added to the bath, with proper controls to prevent a steam explosion.

Hot blackening involves dipping the part into various tanks. The workpiece is usually "dipped" by automated part carriers for transportation between tanks. These tanks contain, in order, alkaline cleaner, water, caustic soda at 140.5 °C (the blackening compound), and finally the sealant, which is usually oil. The caustic soda bonds chemically to the surface of the metal, creating a porous base layer on the part. Oil is then applied to the heated part, which seals it by "sinking" into the applied porous layer. It is the oil that prevents the corrosion of the workpiece. There are many advantages of blackening, mainly:

- blackening can be done in large batches (ideal for small parts),

- no significant dimensional impact (the blacking process creates a layer about a micrometre thick),

- it is far cheaper than similar corrosion protection systems, such as paint and electroplating.

The oldest and most widely used specification for hot black oxide is MIL-DTL-13924, which covers four classes of processes for different substrates. Alternate specifications include AMS 2485, ASTM D769, and ISO 11408.

This is the process used to blacken wire ropes for theatrical applications and flying effects.

## Mid-temperature Black Oxide

Like hot black oxide, mid-temperature black oxide converts the surface of the metal to magnetite ($Fe_3O_4$). However, mid-temperature black oxide blackens at a temperature of 220–245 °F (104–118 °C), significantly less than hot black oxide. This is advantageous because it is below the solution's boiling point, meaning there are no caustic fumes produced.

Since mid-temperature black oxide is most comparable to hot black oxide, it also can meet the military specification MIL-DTL-13924, as well as AMS 2485.

## Cold Black Oxide

Cold black oxide is applied at room temperature. It is not an oxide conversion coating, but rather a deposited copper selenium compound. Cold black oxide offers higher productivity and is convenient for in-house blackening. This coating produces a similar color to the one the oxide conversion does, but tends to rub off easily and offers less abrasion resistance. The application of oil, wax, or lacquer brings the corrosion resistance up to par with the hot and mid-temperature. One application for cold black oxide process would be in tooling and architectural finishing on steel (patina for steel).

## Copper

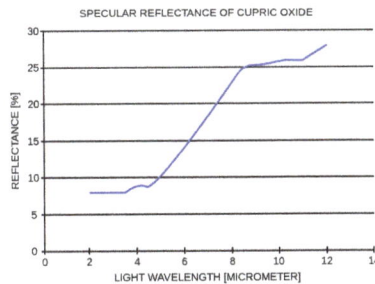

Black oxide for copper, sometimes known by the trade name *Ebonol C*, converts the copper surface to cupric oxide. For the process to work the surface has to have at least 65% copper; for copper surfaces that have less than 90% copper it must first be pretreated with an activating treatment. The finished coating is chemically stable and very adherent. It is stable up to 400 °F (204 °C); above this temperature the coating degrades due to oxidation of the base copper. To increase corrosion resistance, the surface may be oiled, lacquered, or waxed. It is also used as a pre-treatment for painting or enamelling. The surface finish is usually satin, but it can be turned glossy by coating in a clear high-gloss enamel.

On a microscopic scale dendrites form on the surface finish, which trap light and increase absorptivity. Because of this property the coating is used in aerospace, microscopy and other optical applications to minimise light reflection.

In printed circuit boards, the use of black oxide provides better adhesion for the fiberglass laminate layers. The PCB is dipped in a bath containing hydroxide, hypochlorite, and cuprate, which becomes depleted in all three components. This indicates that the black copper oxide comes partially from the cuprate and partially from the PCB copper circuitry. Under microscopic examination, there is no copper(I) oxide layer.

An applicable U.S. military specification is MIL-F-495E.

## Stainless Steel

Hot black oxide for stainless steel is a mixture of caustic, oxidizing, and sulfur salts. It blackens 300 and 400 series, and the precipitation-hardened 17-4 PH stainless steel alloys. The solution can be used on cast iron and mild low-carbon steel. The resulting finish complies with military specification MIL-DTL–13924D Class 4 and offers abrasion resistance. Black oxide finish is used on surgical instruments in light intensive environments to reduce eye fatigue.

Room-temperature blackening for stainless steel occurs by auto-catalytic reaction of copper-selenide depositing on the stainless-steel surface. It offers less abrasion resistance and the same corrosion protection as the hot blackening process. One application for room-temperature blackening is in architectural finishes (patina for stainless steel).

## Zinc

Black oxide for zinc is also known by the trade name *Ebonol Z*. Another product is Ultra-Blak 460, which blackens zinc-plated and galvanized surfaces without using any chrome and zinc die-casts.

## Chromate Conversion Coating

Chromate conversion coating is a type of conversion coating used to passivate steel, aluminum, zinc, cadmium, copper, silver, magnesium, and tin alloys. It is primarily used as a corrosion inhibitor, primer, decorative finish, or to retain electrical conductivity. The process is named after the chromate found in chromic acid, also known as hexavalent chromium, the chemical most widely used in the immersion bath process whereby the coating is applied. However, hexavalent chromium is toxic, thus, highly regulated, so new, non-hexavalent chromium-based processes are becoming more readily available at a commercial level. One alternative contains trivalent chromium. In Europe the RoHS (Restriction of Hazardous Substances) Directive is commonly referred to regarding elimination of hexavalent chromium in electrical and electronic equipment, and the REACH ("Registration, Evaluation, Authorisation and Restriction of Chemicals") Directive to wider applications including chromate conversion coating processes, paint primers and other preparations.

Chromate conversion coatings are commonly applied to everyday items such as hardware and tools, and can usually be recognized by their distinctively iridescent, greenish-yellow color.

Zinc chromate conversion coating on small steel parts.

# Substrates

## Aluminium

Chromate conversion coatings on an aluminium substrate are known by the following terms: *chemical film, yellow iridite,* and the brand names *Iridite* and *Alodine* (in the UK, *Alocrom*). It is also commonly used on aluminum alloy parts in the aircraft industry.

*Iridite NCP* is a non-chromium type of conversion coating for aluminium substrates.

The most commonly referred-to standard for applying chromate conversion coating to aluminium is MIL-DTL-5541 in the US, in the UK it is Def Stan 03/18.

## Magnesium

*Alodine* may also refer to chromate-coating magnesium alloys.

## Phosphate Coatings

Chromate conversion coatings can be applied over phosphate conversion coatings used on ferrous substrates. The process is used to enhance the phosphate coating.

## Steel

Steel and iron cannot be chromated directly. Steel plated with zinc may be chromated.

## Zinc

Chromating is often performed on galvanized parts to make them more durable. The chromate coating acts as paint does, protecting the zinc from white corrosion, thus making the part considerably more durable, depending on the chromate layer's thickness. Steel and iron cannot be chromated directly. Chromating zinc plated steel does not enhance zinc's cathodic protection of the underlying steel from rust.

The protective effect of chromate coatings on zinc is indicated by color, progressing from clear/blue to yellow, gold, olive drab and black. Darker coatings generally provide more corrosion resistance. However, the coating color can also be changed with dyes, so color is not a complete indicator of the process used.

ISO 4520 specifies chromate conversion coatings on electroplated zinc and cadmium coatings. ASTM B633 Type II and III specify zinc plating plus chromate conversion on iron and steel parts.

## Composition

The composition of chromate conversion solutions varies greatly, depending on the material to be coated and the desired effect. Most solution formulae are proprietary.

The widely used Cronak process for zinc and cadmium consists of 5–10 seconds of immersion in a room-temperature solution consisting of 182 g/l sodium dichromate crystals ($Na_2Cr_2O_7 \cdot 2H_2O$) and 6 ml/l concentrated sulfuric acid.

Iridite 14-2, a chromate conversion coating for aluminum, contains chromium(IV) oxide, barium nitrate, sodium silicofluoride and ferricyanide.

Chromate coatings are soft and gelatinous when first applied, but harden and become hydrophobic as they age.

Curing can be accelerated by heating up to 70 °C, but higher temperatures will gradually damage the coating.

# Corrosion Inhibitor

A corrosion inhibitor is a chemical compound that, when added to a liquid or gas, decreases the corrosion rate of a material, typically a metal or an alloy. The effectiveness of a corrosion inhibitor depends on fluid composition, quantity of water, and flow regime. A common mechanism for inhibiting corrosion involves formation of a coating, often a passivation layer, which prevents access of the corrosive substance to the metal. Permanent treatments such as chrome plating are not generally considered inhibitors, however. Instead corrosion inhibitors are *additives* to the fluids that surround the metal or related object.

## Corrosion Inhibitors and their Role

The nature of the corrosive inhibitor depends on (i) the material being protected, which are most commonly metal objects, and (ii) on the corrosive agent(s) to be neutralized.

The corrosive agents are generally oxygen, hydrogen sulfide, and carbon dioxide. Oxygen is generally removed by reductive inhibitors such as amines and hydrazines:

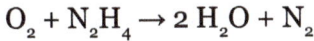

$$O_2 + N_2H_4 \rightarrow 2\,H_2O + N_2$$

In this example, hydrazine converts oxygen, a common corrosive agent, to water, which is generally benign. Related inhibitors of oxygen corrosion are hexamine, phenylenediamine, and dimethylethanolamine, and their derivatives. Antioxidants such as sulfite and ascorbic acid are sometimes used. Some corrosion inhibitors form a passivating coating on the surface by chemisorption. Benzotriazole is one such species used to protect copper. For lubrication, zinc dithiophosphates are common - they deposit sulfide on surfaces.

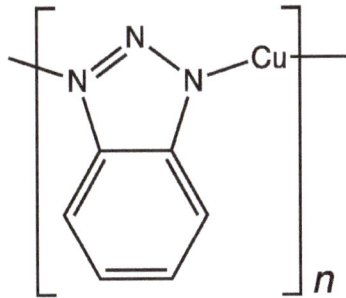

Benzotriazole inhibits corrosion of copper by forming an inert
layer of this polymer on the metal's surface.

The suitability of any given chemical for a task in hand depends on many factors, including their operating temperature.

## Illustrative Applications

- Volatile amines are used in boilers to minimize the effects of acid. In some cases, the amines form a protective film on the steel surface and, at the same time, act as an anodic inhibitor. An inhibitor that acts both in a cathodic and anodic manner is termed a *mixed inhibitor*.

- Benzotriazole inhibits the corrosion and staining of copper surfaces.

- Corrosion inhibitors are often added to paints. A pigment with anticorrosive properties is zinc phosphate. Compounds derived from tannic acid or zinc salts of organonitrogens (e.g. Alcophor 827) can be used together with anticorrosive pigments. Other corrosion inhibitors are Anticor 70, Albaex, Ferrophos, and Molywhite MZAP.

- Antiseptics are used to counter microbial corrosion. Benzalkonium chloride is commonly used in oil field industry.

- In oil refineries, hydrogen sulfide can corrode steels so it is removed often using air and amines by conversion to polysulfides.

## Fuels Industry

Corrosion inhibitors are commonly added to coolants, fuels, hydraulic fluids, boiler water, engine oil, and many other fluids used in industry. For fuels, various corrosion inhibitors can be used. Some components include zinc dithiophosphates.

- DCI-4A, widely used in commercial and military jet fuels, acts also as a lubricity additive. Can be also used for gasolines and other distillate fuels.

- DCI-6A, for motor gasoline and distillate fuels, and for U.S. military fuels (JP-4, JP-5, JP-8)

- DCI-11, for alcohols and gasolines containing oxygenates

- DCI-28, for very low-pH alcohols and gasolines containing oxygenates

- DCI-30, for gasoline and distillate fuels, excellent for pipeline transfers and storage, caustic-resistant

- DMA-4 (solution of alkylaminophosphate in kerosene), for petroleum distillates

## Galvanization

A street lamp in Singapore showing the characteristic spangle of hot-dip galvanization

Galvanization (or galvanizing as it is most commonly called in that industry) is the process of applying a protective zinc coating to steel or iron, to prevent rusting. The most common method is hot-dip galvanizing, in which parts are submerged in a bath of molten zinc.

## Protective Action

Galvanizing protects the base metal in three ways:

- It forms a coating of zinc which, when intact, prevents corrosive substances from reaching the underlying steel or iron.

- The zinc serves as a sacrificial anode so that even if the coating is scratched, the exposed steel will still be protected by the remaining zinc.

- The zinc protects its base metal by corroding before iron. For better results, application of chromates over zinc is also seen as an industrial trend.

## History and Etymology

Galvanized nails

The earliest known example of galvanized iron was encountered by Europeans on 17th-century Indian armor in the Royal Armouries Museum collection. It was named in English via French from the name of Italian scientist Luigi Galvani. Originally in the 19th century, the term "galvanizing" was used to describe the administration of electric shocks; this was also called Faradism. This usage is the origin of the metaphorical use of the verb "galvanize", such as to "galvanize into action" meaning stimulating a complacent person or group to take action.

In modern usage, the term "galvanizing" has largely come to be associated with zinc coatings, to the exclusion of other metals. Galvanic paint, a precursor to hot-dip galvanizing, was patented by Stanislas Sorel, of Paris, in December 1837.

## Methods

Hot-dip galvanizing deposits a thick, robust layer of zinc iron alloys on the surface of a steel item. In the case of automobile bodies, where additional decorative coatings of paint will be applied, a thinner form of galvanizing is applied by electrogalvanizing. The hot-dip process generally does not reduce strength on a measurable scale, with the exception of high-strength steels (>1100 MPa) where hydrogen embrittlement can become a problem. This deficiency is a consideration affecting the manufacture of wire rope and other highly-stressed products.

The protection provided by hot-dip galvanizing is insufficient for products that will be constantly exposed to corrosive materials such as acids, including acid rain in outdoor uses. For these applications, more expensive stainless steel is preferred. Some nails made today are galvanized. Nonetheless, electroplating is used on its own for many outdoor applications because it is cheaper than hot-dip zinc coating and looks good when new. Another reason not to use hot-dip zinc coating is that for bolts and nuts of size M10 (US 3/8") or smaller, the thick hot-dipped coating fills in too much of the threads, which reduces strength (because the dimension of the steel prior to coating must be reduced for the fasteners to fit together). This means that for cars, bicycles, and many other light mechanical products, the practical alternative to electroplating bolts and nuts is not hot-dip zinc coating, but making the fasteners from stainless steel.

The size of crystallites in galvanized coatings is a visible and aesthetic feature, known as "spangle". By varying the number of particles added for heterogeneous nucleation and the rate of cooling in a hot-dip process, the spangle can be adjusted from an apparently uniform surface (crystallites too small to see with the naked eye) to grains several centimetres wide. Visible crystallites are rare in other engineering materials, even though they are usually present.

Thermal diffusion galvanizing, or Sherardizing, provides a zinc diffusion coating on iron- or copper-based materials. Parts and zinc powder are tumbled in a sealed rotating drum. Around 300 °C (572 °F), zinc will diffuse into the substrate to form a zinc alloy. The advance surface preparation of the goods can be carried out by shot blasting. The process is also known as "dry galvanizing", because no liquids are involved; this can avoid possible problems caused by hydrogen embrittlement. The dull-grey crystal structure of the zinc diffusion coating has a good adhesion to paint, powder coatings, or rubber. It is a preferred method for coating small, complex-shaped metals, and for smoothing rough surfaces on items formed with sintered metal.

## Eventual Corrosion

Rusted corrugated steel roof

Although galvanizing will inhibit attack of the underlying steel, rusting will be inevitable after some decades of exposure to weather, especially if exposed to acidic conditions. For example, corrugated iron sheet roofing will start to degrade within a few years despite the protective action of the zinc coating. Marine and salty environments also lower the lifetime of galvanized iron because the high electrical conductivity of sea water increases the rate of corrosion, primarily through converting the solid zinc to soluble zinc chloride which simply washes away. Galvanized car frames exemplify this; they corrode much faster in cold environments due to road salt, though they will last longer than unprotected steel.

Galvanized steel can last for many decades if other supplementary measures are maintained, such as paint coatings and additional sacrificial anodes. The rate of corrosion in non-salty environments is caused mainly by levels of sulfur dioxide in the air. In the most benign natural environments, such as inland low population areas, galvanized steel can last without rust for over 100 years.

## Galvanized Construction Steel

This is the most common use for galvanized metal, and hundreds of thousands of tons of steel products are galvanized annually worldwide. In developed countries most larger cities have several galvanizing factories, and many items of steel manufacture are galvanized for protection. Typically these include: street furniture, building frameworks, balconies, verandahs, staircases, ladders, walkways, and more. SGCC hot dip galvanized steel is also used for making steel frames as a basic construction material for steel frame buildings.

## Galvanized Piping

In the early 20th century, galvanized piping replaced previously-used cast iron and lead in cold-water plumbing. Typically, galvanized piping rusts from the inside out, building up layers of plaque on the inside of the piping, causing both water pressure problems and eventual pipe failure. These plaques can flake off, leading to visible impurities in water and a slight metallic taste. The life expectancy of galvanized piping is about 70 years, but it may vary by region due to impurities in the water supply and the proximity of electrical grids for which interior piping acts as a pathway (the flow of electricity can accelerate chemical corrosion). Pipe longevity also depends on the thickness of zinc in the original galvanizing, which ranges on a scale from G40 to G210, and whether the pipe was galvanized on both the inside and outside, or just the outside.

Since World War II, copper and plastic piping have replaced galvanized piping for interior drinking water service, but galvanized steel pipes are still used in outdoor applications requiring steel's superior mechanical strength. The use of galvanized pipes lends some truth to the urban myth that water purity in outdoor water faucets is lower, but the actual impurities (iron, zinc, calcium) are harmless.

The presence of galvanized piping detracts from the appraised value of housing stock because piping can fail, increasing the risk of water damage. Galvanized piping will eventually need to be replaced if housing stock is to outlast a 50 to 70 year life expectancy, and some jurisdictions require galvanized piping to be replaced before sale. One option to extend the life expectancy of existing galvanized piping is to line it with an epoxy resin.

## Electrogalvanization

Electrogalvanizing is a process in which a layer of zinc is bonded to steel in order to protect against corrosion. The process involves electroplating, running a current of electricity through a saline/zinc solution with a zinc anode and steel conductor. Zinc electroplating maintains a dominant position among other electroplating process options, based upon electroplated tonnage per annum. According to the International Zinc Association, more than 5 million tons are used yearly for both Hot Dip Galvanizing and Electroplating. The Plating of Zinc was developed at the beginning of the 20th century. At that time, the electrolyte was cyanide based. A significant innovation occurred in the 1960s, with the introduction of the first acid chloride based electrolyte. The 1980s saw a return to alkaline electrolytes, only this time, without the use of cyanide. The most commonly used electrogalvanized cold rolled steel is SECC steel. Compared to hot dip galvanizing, electroplated zinc offers these significant advantages:

- Lower thickness deposits to achieve comparable performance

- Broader conversion coating availability for increased performance and colour options

- Brighter, more aesthetically appealing, deposits

## History

Zinc plating was developed and continues to evolve, to meet the most challenging corrosion protection, temperature and wear resistance requirements. Electroplating of zinc was invented in 1800 but the first bright deposits were not obtained until the early 1930s with the alkaline cyanide electrolyte. Much later, in 1966, the use of acid chloride baths improved the brightness even greater. The latest modern development occurred in the 1980s, with the new generation of alkaline, cyanide-free zinc. Recent European Union directives (ELV/RoHS/WEEE) prohibit automotive, other original equipment manufacturers (OEM) and electrical and electronic equipment manufacturers from using hexavalent chromium (CrVI). These directives combined with increased performance requirements by the OEM, has led to an increase in the use of alkaline zinc, zinc alloys and high performance trivalent passionate conversion coatings.

## Processes

The corrosion protection afforded by the electrodeposited zinc layer is primarily due to the anodic potential dissolution of zinc versus iron (the substrate in most cases). Zinc acts as a sacrificial anode for protecting the iron (steel). While steel is close to $E_{SCE}$= -400 mV (the potential refers to the standard Saturated calomel electrode (SCE), depending on the alloy composition, electroplated zinc is much more anodic with $E_{SCE}=$ -980 mV. Steel is preserved from corrosion by cathodic protection. Conversion coatings (hexavalent chromium (CrVI) or trivalent chromium (CrIII) depending upon OEM requirements) are applied to drastically enhance the corrosion protection by building an additional inhibiting layer of Chromium and Zinc hydroxides. These oxide films range in thickness from 10 nm for the thinnest blue/clear passivates to 4 µm for the thickest black chromates.

Additionally, electroplated zinc articles may receive a topcoat to further enhance corrosion protection and friction performance.

The modern electrolytes are both alkaline and acidic:

# Alkaline Electrolytes

## Cyanide Electrolytes

Contain sodium sulphate and sodium hydroxide (NaOH). All of them utilize proprietary brightening agents. Zinc is soluble as a cyanide complex $Na_2Zn(CN)_4$ and as a zincate $Na_2Zn(OH)_4$. Quality control of such electrolytes requires the regular analysis of Zn, NaOH and NaCN. The ratio of NaCN : Zn can vary between 2 and 3 depending upon the bath temperature and desired deposit brightness level.The following chart illustrates the typical cyanide electrolyte options used to plate at room temperature:

| Cyanide bath composition | | | |
|---|---|---|---|
| | **Zinc** | **Sodium hydroxide** | **Sodium cyanide** |
| **Low cyanide** | 6-10 g/L (0.8-1.3 oz/gal) | 75-90 g/L (10-12 oz/gal) | 10-20 g/L 1.3-2.7 oz/gal) |
| **Mid cyanide** | 15-20 g/L (2.0-2.7 oz/gal) | 75-90 g/L (10-12 oz/gal) | 25-45 g/L (3.4-6.0 oz/gal) |
| **High cyanide** | 25-35 g/L (3.4-4.7 oz/gal) | 75-90 g/L (10-12 oz/gal) | 80-100 g/L (10.70- 13.4 oz/gal) |

# Alkaline Non-cyanide Electrolytes

Contain zinc and sodium hydroxide. Most of them are brightened by proprietary addition agents similar to those used in cyanide baths. The addition of quaternary amine additives contribute to the improved metal distribution between high and low current density areas. Depending upon the desired performance, the electroplater can select the highest zinc content for increased productivity or lower zinc content for a better throwing power (into low current density areas). For ideal metal distribution, Zn metal evolutes between 6-14 g/L (0.8-1.9 oz/gal) and NaOH at 120 g/L (16 oz/gal). But for the highest productivity, Zn metal is between 14-25 g/L (1.9-3.4 oz/gal) and

NaOH remains at 120 g/L (16 oz/gal). Alkaline Non Cyanide Zinc Process contains lower concentration zinc metal concentration 6-14 g/L (0.8-1.9 oz/gal) or higher zinc metal concentration 14-25 g/L (1.9-3.4 oz/gal) provides superior plate distribution from high current density to low current density or throwing power when compared to any acidic baths such as chloride based (Low ammonium chloride, Potassium chloride / Ammonium Chloride) - or (non-ammonium chloride, potassium chloride/Boric acid) or sulfate baths.

## Acidic Electrolytes

### High Speed Electrolytes

Dedicated to plating at high speed in plants where the shortest plating time is critical (i.e. steel coil or pipe that runs at up to 200 m/min. The baths contain zinc sulfate and chloride to the maximum solubility level. Boric acid may be used as a pH buffer and to reduce the burning effect at high current densities. These baths contain very few grain refiners. If one is utilized, it may be sodium saccharine.

### Traditional Electrolytes

Initially based on ammonium chloride, options today include ammonium, potassium or mixed ammonium/potassium electrolytes. The chosen content of zinc depends on the required productivity and part configuration. High zinc improves the bath's efficiency (plating speed), while lower levels improve the bath's ability to throw into low current densities. Typically, the Zn metal level varies between 20 and 50 g/L (2.7-6.7 oz/gal). The pH varies between 4.8 and 5.8 units. The following chart illustrates a typical all potassium chloride bath composition:

| Traditional acid bath composition | |
|---|---|
| **Parameters** | **Value in g/L (oz/gal)** |
| Zinc | 40 g/l (5.4 oz/gal) |
| Total chloride | 125 g/l (16.8 oz/gal) |
| Anhydrous zinc chloride | 80 g/l (10.7 oz/gal) |
| Potassium chloride | 180 g/l (24.1 oz/gal) |
| Boric acid | 25 g/l (3.4 oz/gal) |

Typical grain refiners include low soluble ketones and aldehydes. These brightening agents must be dissolved in alcohol or in hydrotrope. The resultant molecules are co-deposited with the zinc to produce a slightly leveled, very bright deposit. The bright deposit has also been shown to decrease chromate/passivate receptivity, however. The result is a reduction in the corrosion protection afforded.

# Hot-dip Galvanization

Crystalline surface of a hot-dip galvanized handrail.

Hot-dip galvanization is a form of galvanization. It is the process of coating iron and steel with zinc, which alloys with the surface of the base metal when immersing the metal in a bath of molten zinc at a temperature of around 840 °F (449 °C). When exposed to the atmosphere, the pure zinc (Zn) reacts with oxygen ($O_2$) to form zinc oxide (ZnO), which further reacts with carbon dioxide ($CO_2$) to form zinc carbonate ($ZnCO_3$), a usually dull grey, fairly strong material that protects the steel underneath from further corrosion in many circumstances. Galvanized steel is widely used in applications where corrosion resistance is needed without the cost of stainless steel, and is considered superior in terms of cost and life-cycle. It can be identified by the crystallization patterning on the surface (often called a "spangle").

Galvanized steel can be welded; however, one must exercise caution around the resulting toxic zinc fumes. Galvanized steel is suitable for high-temperature applications of up to 392 °F (200 °C). The use of galvanized steel at temperatures above this will result in peeling of the zinc at the *inter metallic* layer. Electrogalvanized sheet steel is often used in automotive manufacturing to enhance the corrosion performance of exterior body panels; this is, however, a completely different process which tends to achieve lower coating thicknesses of zinc.

Like other corrosion protection systems, galvanizing protects steel by acting as a barrier between steel and the atmosphere. However, zinc is a more electropositive (active) metal in comparison to steel. This is a unique characteristic for galvanizing, which means that when a galvanized coating is damaged and steel is exposed to the atmosphere, zinc can continue to protect steel through galvanic corrosion (often within an annulus of 5 mm, above which electron transfer rate decreases).

## Process

The process of hot-dip galvanizing results in a metallurgical bond between zinc and steel with a series of distinct iron-zinc alloys. The resulting coated steel can be used in much the same way as uncoated.

A typical hot-dip galvanizing line operates as follows:

- Steel is cleaned using a caustic solution. This removes oil/grease, dirt, and paint.

- The caustic cleaning solution is rinsed off.

- The steel is pickled in an acidic solution to remove mill scale.

- The pickling solution is rinsed off.

- A flux, often zinc ammonium chloride is applied to the steel to inhibit oxidation of the cleaned surface upon exposure to air. The flux is allowed to dry on the steel and aids in the process of the liquid zinc wetting and adhering to the steel.

- The steel is dipped into the molten zinc bath and held there until the temperature of the steel equilibrates with that of the bath.

- The steel is cooled in a quench tank to reduce its temperature and inhibit undesirable reactions of the newly formed coating with the atmosphere.

Lead is often added to the molten zinc bath to improve the fluidity of the bath (thus limiting excess zinc on the dipped product by improved drainage properties), helps prevent floating dross, makes dross recycling easier and protects the Pilling kettle from uneven heat distribution from the burners. Lead is either added to primary Z1 grade zinc or already contained in used secondary zinc. A third, declining method is to use low Z5 grade zinc.

Steel strip can be hot-dip galvanized in a continuous line. Hot-dip galvanized steel strip (also sometimes loosely referred to as galvanized iron) is extensively used for applications requiring the strength of steel combined with the resistance to corrosion of zinc. roofing and walling, safety barriers, handrails, consumer appliances and automotive body parts. One common use is in metal pails. Galvanised steel is also used in most heating and cooling duct systems in buildings.

Individual metal articles, such as steel girders or wrought iron gates, can be hot-dip galvanized by a process called batch galvanizing. Other modern techniques have largely replaced hot-dip for these sorts of roles. This includes *electrogalvanizing*, which deposits the layer of zinc from an aqueous electrolyte by electroplating, forming a thinner and much stronger bond.

## History

In 1742, French chemist Paul Jacques Malouin described a method of coating iron by dipping it in molten zinc in a presentation to the French Royal Academy.

In 1772 Luigi Galvani (Italy), galvanizing's namesake, discovered the electrochemical process that takes place bettween metals during an experiment with frog legs.

In 1801 Alessandro Volta furthered the research on galvanizing when he discovered the electro-potential between two metals, creating a corrosion cell.

In 1836, French chemist Stanislas Sorel obtained a patent for a method of coating iron with zinc, after first cleaning it with 9% sulfuric acid ($H_2SO_4$) and fluxing it with ammonium chloride ($NH_4Cl$).

## Specification

A hot-dip galvanized coating is relatively easier and cheaper to specify than an organic paint coating of equivalent corrosion protection performance. The British, European and International standard for hot-dip galvanizing is BS EN ISO 1461 which specifies a minimum coating thickness to be applied to steel in relation to the steels section thickness e.g. a steel fabrication with a section size thicker than 6 mm shall have a minimum galvanized coating thickness of 85 µm.

Further performance and design information for galvanizing can be found in BS EN ISO 14713-1 and BS EN ISO 14713-2. The durability performance of a galvanized coating depends solely on the corrosion rate of the environment in which it is placed. Corrosion rates for different environments can be found in BS EN ISO 14713-1 where typical corrosion rates are given with a description of the environment in which the steel would be used.

## Anodic Protection

Anodic protection (AP) is a technique to control the corrosion of a metal surface by making it the anode of an electrochemical cell and controlling the electrode potential in a zone where the metal is passive.

AP is used to protect metals that exhibit passivation in environments whereby the current density in the freely corroding state is significantly higher than the current density in the passive state over a wide range of potentials.

Anodic protection is used for carbon steel storage tanks containing extreme pH environments including concentrated sulfuric acid and 50 percent caustic soda where cathodic protection is not suitable due to very high current requirements.

An anodic protection system includes an external power supply connected to auxiliary cathodes and controlled by a feedback signal from one or more reference electrodes. Careful design and control is required when using anodic protection for several reasons, including excessive current when passivation is lost or unstable, leading to possible accelerated corrosion.

## Cathodic Protection

Cathodic protection (CP) is a technique used to control the corrosion of a metal surface by making it the cathode of an electrochemical cell. A simple method of protection connects the metal to be protected to a more easily corroded "sacrificial metal" to act as the anode. The sacrificial metal then corrodes instead of the protected metal. For structures such as long pipelines, where passive galvanic cathodic protection is not adequate, an external DC electrical power source is used to provide sufficient current.

Zinc sacrificial anode *(rounded object)* screwed to the underside of the hull of a small boat.

Cathodic protection systems protect a wide range of metallic structures in various environments. Common applications are: steel water or fuel pipelines and steel storage tanks such as home water heaters; steel pier piles; ship and boat hulls; offshore oil platforms and onshore oil well casings; offshore wind farm foundations and metal reinforcement bars in concrete buildings and structures. Another common application is in galvanized steel, in which a sacrificial coating of zinc on steel parts protects them from rust.

Cathodic protection can, in some cases, prevent stress corrosion cracking.

### History

Cathodic protection was first described by Sir Humphry Davy in a series of papers presented to the Royal Society in London in 1824. The first application was to the HMS Samarang in 1824. Sacrificial anodes made from iron attached to the copper sheath of the hull below the waterline dramatically reduced the corrosion rate of the copper. However, a side effect of the cathodic protection was to increase marine growth. Copper, when corroding, releases copper ions which have an anti-fouling effect. Since excess marine growth affected the performance of the ship, the Royal Navy decided that it was better to allow the copper to corrode and have the benefit of reduced marine growth, so cathodic protection was not used further.

Davy was assisted in his experiments by his pupil Michael Faraday, who continued his research after Davy's death. In 1834, Faraday discovered the quantitative connection between corrosion weight loss and electric current and thus laid the foundation for the future application of cathodic protection.

Thomas Edison experimented with impressed current cathodic protection on ships in 1890, but was unsuccessful due to the lack of a suitable current source and anode materials. It would be 100 years after Davy's experiment before cathodic protection was used widely on oil pipelines in the United States — cathodic protection was applied to steel gas pipelines beginning in 1928 and more widely in the 1930s.

## Galvanic

In the application of *passive* cathodic protection, a *galvanic anode*, a piece of a more electrochemically "active" metal, is attached to the vulnerable metal surface where it is exposed to an electrolyte. Galvanic anodes are selected because they have a more "active" voltage (more negative electrode potential) than the metal of the target structure (typically steel). For effective cathodic protection, the potential of the steel surface is polarized (pushed) more negative until the surface has a uniform potential. At that stage, the driving force for the corrosion reaction with the protected surface is removed. The galvanic anode continues to corrode, consuming the anode material until eventually it must be replaced. Polarization of the target structure is caused by the electron flow from the anode to the cathode, so the two metals must have a good electrically conductive contact. The driving force for the cathodic protection current is the difference in electrode potential between the anode and the cathode.

Galvanic or sacrificial anodes are made in various shapes and sizes using alloys of zinc, magnesium and aluminium. ASTM International publishes standards on the composition and manufacturing of galvanic anodes.

In order for galvanic cathodic protection to work, the anode must possess a lower (that is, more negative) electrode potential than that of the cathode (the target structure to be protected). The table below shows a simplified galvanic series which is used to select the anode metal. The anode must be chosen from a material that is lower on the list than the material to be protected.

| Metal | Potential with respect to a $Cu:CuSO_4$ reference electrode in neutral pH environment (volts) |
|---|---|
| Carbon, Graphite, Coke | +0.3 |

| Platinum | 0 to -0.1 |
|---|---|
| Mill scale on Steel | -0.2 |
| High Silicon Cast Iron | -0.2 |
| Copper, brass, bronze | -0.2 |
| Mild steel in concrete | -0.2 |
| Lead | -0.5 |
| Cast iron (not graphitized) | -0.5 |
| Mild steel (rusted) | -0.2 to -0.5 |
| Mild steel (clean) | -0.5 to -0.8 |
| Commercially pure aluminium | -0.8 |
| Aluminium alloy (5% zinc) | -1.05 |
| Zinc | -1.1 |
| Magnesium Alloy (6% Al, 3% Zn, 0.15% Mn) | -1.6 |
| Commercially Pure Magnesium | -1.75 |

## Impressed Current Systems

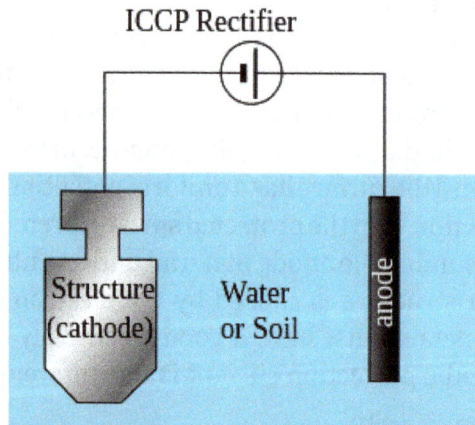

Simple impressed current cathodic protection system.
A source of DC electric current is used to help drive the protective electrochemical reaction.

For larger structures, or where electrolyte resistivity is high, galvanic anodes cannot economically deliver enough current to provide protection. In these cases, *impressed current cathodic protection* (ICCP) systems are used. These consist of anodes connected to a DC power source, often a transformer-rectifier connected to AC power. In the absence of an AC supply, alternative power sources may be used, such as solar panels, wind power or gas powered thermoelectric generators.

Anodes for ICCP systems are available in a variety of shapes and sizes. Common anodes are tubular and solid rod shapes or continuous ribbons of various materials. These include high silicon cast iron, graphite, mixed metal oxide, platinum and niobium coated wire and other materials.

For pipelines, anodes are arranged in groundbeds either distributed or in a deep vertical hole depending on several design and field condition factors including current distribution requirements.

Cathodic protection transformer-rectifier units are often custom manufactured and equipped with a variety of features, including remote monitoring and control, integral current interrupters and various type of electrical enclosures. The output DC negative terminal is connected to the structure to be protected by the cathodic protection system. The rectifier output DC positive cable is connected to the anodes. The AC power cable is connected to the rectifier input terminals.

The output of the ICCP system should be optimised to provide enough current to provide protection to the target structure. Some cathodic protection transformer-rectifier units are designed with taps on the transformer windings and jumper terminals to select the voltage output of the ICCP system. Cathodic protection transformer-rectifier units for water tanks and used in other applications are made with solid state circuits to automatically adjust the operating voltage to maintain the optimum current output or structure-to-electrolyte potential. Analog or digital meters are often installed to show the operating voltage (DC and sometime AC) and current output. For shore structures and other large complex target structures, ICCP system are often designed with multiple independent zones of anodes with separate cathodic protection transformer-rectifier circuits.

## Applications

## Pipelines

An air cooled cathodic protection rectifier connected to a pipeline.

Hazardous product pipelines are routinely protected by a coating supplemented with cathodic protection. An impressed current cathodic protection system (ICCP) for a pipeline consists of a DC power source, often an AC powered transformer rectifier and an anode, or array of anodes buried in the ground (the anode groundbed).

The DC power source would typically have a DC output of up to 50 amperes and 50 volts, but this depends on several factors, such as the size of the pipeline and coating quality. The positive DC output terminal would be connected via cables to the anode array, while another cable would connect the negative terminal of the rectifier to the pipeline, preferably through junction boxes to allow measurements to be taken.

Anodes can be installed in a groundbed consisting of a vertical hole backfilled with conductive coke (a material that improves the performance and life of the anodes) or laid in a prepared trench, surrounded by conductive coke and backfilled. The choice of groundbed type and size depends on the application, location and soil resistivity.

The DC cathodic protection current is then adjusted to the optimum level after conducting various tests including measurements of pipe-to-soil potentials or electrode potential.

It is sometimes more economically viable to protect a pipeline using galvanic (sacrificial) anodes. This is often the case on smaller diameter pipelines of limited length. Galvanic anodes rely on the galvanic series potentials of the metals to drive cathodic protection current from the anode to the structure being protected.

Water pipelines of various pipe materials are also provided with cathodic protection where owners determine the cost is reasonable for the expected pipeline service life extension attributed to the application of cathodic protection.

## Ships and Boats

The white patches visible on the ship's hull are zinc block sacrificial anodes.

Cathodic protection on ships is often implemented by galvanic anodes attached to the hull and ICCP for larger vessels. Since ships are regularly removed from the water for inspections and maintenance, it is a simple task to replace the galvanic anodes.

Galvanic anodes are generally shaped to reduced drag in the water and fitted flush to the hull to also try to minimize drag.

Smaller vessels, with non-metallic hulls, such as yachts, are equipped with galvanic anodes to protect areas such as outboard motors. As with all galvanic cathodic protection, this application relies on a solid electrical connection between the anode and the item to be protected.

For ICCP on ships, the anodes are usually constructed of a relatively inert material such as platinised titanium. A DC power supply is provided within the ship and the anodes mounted on the outside of the hull. The anode cables are introduced into the ship via a compression seal fitting and routed to the DC power source. The negative cable from the power supply is simply attached to the hull to complete the circuit. Ship ICCP anodes are flush-mounted, minimizing the effects of drag on the ship, and located a minimum 5 ft below the light load line in an area to avoid mechanical damage. The current density required for protection is a function of velocity and considered when selecting the current capacity and location of anode placement on the hull.

Some ships may require specialist treatment, for example aluminium hulls with steel fixtures will create an electrochemical cell where the aluminium hull can act as a galvanic anode and corrosion is enhanced. In cases like this, aluminium or zinc galvanic anodes can be used to offset the potential difference between the aluminium hull and the steel fixture. If the steel fixtures are large, several galvanic anodes may be required, or even a small ICCP system.

## Marine

Marine cathodic protection covers many areas, jetties, harbors, offshore structures. The variety of different types of structure leads to a variety of systems to provide protection. Galvanic anodes are favored, but ICCP can also often be used. Because of the wide variety of structure geometry, composition, and architecture, specialized firms are often required to engineer structure-specific cathodic protection systems. Sometimes marine structures require retroactive modification to be effectively protected.

## Steel in Concrete

The application to concrete reinforcement is slightly different in that the anodes and reference electrodes are usually embedded in the concrete at the time of construction when the concrete is being poured. The usual technique for concrete buildings, bridges and similar structures is to use ICCP, but there are systems available that use the principle of galvanic cathodic protection as well, although in the UK at least, the use of galvanic anodes for atmospherically exposed reinforced concrete structures is considered experimental.

For ICCP, the principle is the same as any other ICCP system. However, in a typical atmospherically exposed concrete structure such as a bridge, there will be many more anodes distributed through the structure as opposed to an array of anodes as used on a pipeline. This makes for a more complicated system and usually an automatically controlled DC power source is used, possibly with an option for remote monitoring and operation. For buried or submerged structures, the treatment is similar to that of any other buried or submerged structure.

Galvanic systems offer the advantage of being easier to retrofit and do not need any control systems as ICCP does.

For pipelines constructed from pre-stressed concrete cylinder pipe (PCCP), the techniques used for cathodic protection are generally as for steel pipelines except that the applied potential must be limited to prevent damage to the prestressing wire.

The steel wire in a PCCP pipeline is stressed to the point that any corrosion of the wire can result in failure. An additional problem is that any excessive hydrogen ions as a result of an excessively negative potential can cause hydrogen embrittlement of the wire, also resulting in failure. The failure of too many wires will result in catastrophic failure of the PCCP. To implement ICCP therefore requires very careful control to ensure satisfactory protection. A simpler option is to use galvanic anodes, which are self-limiting and need no control.

## Internal Cathodic Protection

Vessels, pipelines and tanks which are used to store or transport liquids can also be protected from corrosion on their internal surfaces by the use of cathodic protection. ICCP and galvanic systems can be used. A common application of internal cathodic protection is water storage tanks and power plant shell and tube heat exchangers.

## Galvanized Steel

Galvanizing generally refers to hot-dip galvanizing which is a way of coating steel with a layer of metallic zinc or tin. Galvanized coatings are quite durable in most environments because they combine the barrier properties of a coating with some of the benefits of cathodic protection. If the zinc coating is scratched or otherwise locally damaged and steel is exposed, the surrounding areas of zinc coating form a galvanic cell with the exposed steel and protect it from corrosion. This is a form of localized cathodic protection - the zinc acts as a sacrificial anode.

Galvanizing, while using the electrochemical principle of cathodic protection, is not actually cathodic protection. Cathodic protection requires the anode to be separate from the metal surface to be protected, with an ionic connection through the electrolyte and an electron connection through a connecting cable, bolt or similar. This means that any area of the protected structure within the electrolyte can be protected, whereas in the case of galvanizing, only areas very close to the zinc are protected. Hence, a larger area of bare steel would only be protected around the edges.

## Automobiles

Several companies market electronic devices claiming to mitigate corrosion for automobiles and trucks. Corrosion control professionals find they do not work. There is no peer reviewed scientific testing and validation supporting the use of the devices. In 1996 the

FTC ordered David McCready, a person that sold devices claiming to protect cars from corrosion, to pay restitution and banned the names "Rust Buster" and "Rust Evader."

## Testing

Electrode potential is measured with reference electrodes. Copper-copper sulphate electrodes are used for structures in contact with soil or fresh water. Silver/silver chloride/seawater electrodes or pure zinc electrodes are used for seawater applications. The methods are described in EN 13509:2003 and NACE TM0497 along with the sources of error in the voltage that appears on the display of the meter. Interpretation of electrode potential measurements to determine the potential at the interface between the anode of the corrosion cell and the electrolyte requires training and cannot be expected to match the accuracy of measurements done in laboratory work.

## Problems

### Production of Hydrogen

A side effect of improperly applied cathodic protection is the production of atomic hydrogen, leading to its absorption in the protected metal and subsequent hydrogen embrittlement of welds and materials with high hardness. Under normal conditions, the atomic hydrogen will combine at the metal surface to create hydrogen gas, which cannot penetrate the metal. Hydrogen atoms, however, are small enough to pass through the crystalline steel structure, and lead in some cases to hydrogen embrittlement.

### Cathodic Disbonding

This is a process of disbondment of protective coatings from the protected structure (cathode) due to the formation of hydrogen ions over the surface of the protected material (cathode). Disbonding can be exacerbated by an increase in alkali ions and an increase in cathodic polarization. The degree of disbonding is also reliant on the type of coating, with some coatings affected more than others. Cathodic protection systems should be operated so that the structure does not become excessively polarized, since this also promotes disbonding due to excessively negative potentials. Cathodic disbonding occurs rapidly in pipelines that contain hot fluids because the process is accelerated by heat flow.

### Cathodic Shielding

Effectiveness of cathodic protection (CP) systems on steel pipelines can be impaired by the use of solid film backed dielectric coatings such as polyethylene tapes, shrinkable pipeline sleeves, and factory applied single or multiple solid film coatings. This phenomenon occurs because of the high electrical resistivity of these film backings. Protective electric current from the cathodic protection system is blocked or shielded from reaching the underlying metal by the highly resistive film backing. Cathodic shielding was first defined in the 1980s as being a problem, and technical papers on the subject have been regularly published since then.

A 1999 report concerning a 20,600 bbl (3,280 m³) spill from a Saskatchewan crude oil line contains an excellent definition of the cathodic shielding problem:

> "The triple situation of disbondment of the (corrosion) coating, the dielectric nature of the coating and the unique electrochemical environment established under the exterior coating, which acts as a shield to the electrical CP current, is referred to as CP shielding. The combination of tenting and disbondment permits a corrosive environment around the outside of the pipe to enter into the void between the exterior coating and the pipe surface. With the development of this CP shielding phenomenon, impressed current from the CP system cannot access exposed metal under the exterior coating to protect the pipe surface from the consequences of an aggressive corrosive environment. The CP shielding phenomenon induces changes in the potential gradient of the CP system across the exterior coating, which are further pronounced in areas of insufficient or sub-standard CP current emanating from the pipeline's CP system. This produces an area on the pipeline of insufficient CP defense against metal loss aggravated by an exterior corrosive environment."

Cathodic shielding is referenced in a number of the standards listed below. Newly issued USDOT regulation Title 49 CFR 192.112, in the section for *Additional design requirements for steel pipe using alternative maximum allowable operating pressure* requires that "The pipe must be protected against external corrosion by a non-shielding coating". Also, the NACE SP0169:2007 standard defines shielding in section 2, cautions against the use of materials that create electrical shielding in section 4.2.3, cautions against use of external coatings that create electrical shielding in section 5.1.2.3, and instructs readers to take 'appropriate action' when the effects of electrical shielding of cathodic protection current are detected on an operating pipeline in section 10.9.

## Standards

- 49 CFR 192.451 - Requirements for Corrosion Control - Transportation of natural and other gas by pipeline: US minimum federal safety standards

- 49 CFR 195.551 - Requirements for Corrosion Control - Transportation of hazardous liquids by pipelines: US minimum federal safety standards

- AS 2832.4 - Australian Standard for Cathodic Protection

- ASME B31Q 0001-0191

- ASTM G 8, G 42 - Evaluating Cathodic Disbondment resistance of coatings

- DNV-RP-B401 - Cathodic Protection Design - Det Norske Veritas

- EN 12068:1999 - Cathodic protection. External organic coatings for the corrosion protection of buried or immersed steel pipelines used in conjunction with cathodic protection. Tapes and shrinkable materials

- EN 12473:2000 - General principles of cathodic protection in sea water

- EN 12474:2001 - Cathodic protection for submarine pipelines

- EN 12495:2000 - Cathodic protection for fixed steel offshore structures

- EN 12499.2003 - Internal cathodic protection of metallic structures

- EN 12696:2012 - Cathodic protection of steel in concrete

- EN 12954:2001 - Cathodic protection of buried or immersed metallic structures. General principles and application for pipelines

- EN 13173:2001 - Cathodic protection for steel offshore floating structures

- EN 13174:2001 - Cathodic protection for "Harbour Installations".

- EN 13509:2003 - Cathodic protection measurement techniques

- EN 13636:2004 - Cathodic protection of buried metallic tanks and related piping

- EN 14505:2005 - Cathodic protection of complex structures

- EN 15112:2006 - External cathodic protection of well casing

- EN 15280-2013 - Evaluation of a.c. corrosion likelihood of buried pipelines

- EN 50162:2004 - Protection against corrosion by stray current from direct current systems

- BS 7361-1:1991 - Cathodic Protection

- NACE SP0169:2013 - Control of External Corrosion on Underground or Submerged Metallic Piping Systems

- NACE TM 0497 - Measurement Techniques Related to Criteria for Cathodic Protection on Underground or Submerged Metallic Piping Systems

## Coating

A coating is a covering that is applied to the surface of an object, usually referred to as the substrate. The purpose of applying the coating may be decorative, functional, or both. The coating itself may be an all-over coating, completely covering the substrate, or it may only cover parts of the substrate. An example of all of these types of coating is a product label on many drinks bottles- one side has an all-over functional coating (the adhesive) and the other side has one or more decorative coatings in an appropriate pattern (the printing) to form the words and images.

Paints and lacquers are coatings that mostly have dual uses of protecting the substrate and being decorative, although some artists paints are only for decoration, and the paint on large industrial pipes is presumably only for the function of preventing corrosion.

Functional coatings may be applied to change the surface properties of the substrate, such as adhesion, wetability, corrosion resistance, or wear resistance. In other cases, e.g. semiconductor device fabrication (where the substrate is a wafer), the coating adds a completely new property such as a magnetic response or electrical conductivity and forms an essential part of the finished product.

A major consideration for most coating processes is that the coating is to be applied at a controlled thickness, and a number of different processes are in use to achieve this control, ranging from a simple brush for painting a wall, to some very expensive machinery applying coatings in the electronics industry. A further consideration for 'non-all-over' coatings is that control is needed as to where the coating is to be applied. A number of these non-all-over coating processes are printing processes.

Many industrial coating processes involve the application of a thin film of functional material to a substrate, such as paper, fabric, film, foil, or sheet stock. If the substrate starts and ends the process wound up in a roll, the process may be termed "roll-to-roll" or "web-based" coating. A roll of substrate, when wound through the coating machine, is typically called a web.

Coatings may be applied as liquids, gases or solids.

## Functions of Coatings

- Adhesive – adhesive tape, pressure-sensitive labels, iron-on fabric
- Changing adhesion properties
  - » Non-stick PTFE coated- cooking pans
  - » Release coatings e.g. silicone-coated release liners for many self-adhesive products
  - » primers encourage subsequent coatings to adhere well (also sometimes have anti-corrosive properties)
- Optical coatings
  - » Reflective coatings for mirrors

- » Anti-reflective coatings e.g. on spectacles
- » UV- absorbent coatings for protection of eyes or increasing the life of the substrate
- » Tinted as used in some coloured lighting, tinted glazing, or sunglasses
- Catalytic e.g. some self-cleaning glass
- Light-sensitive as previously used to make photographic film
- Protective coatings
  - » Most surface coatings or paints are to some extent protecting the substrate e.g.
    - ◇ Sealing and waterproofing wood
    - ◇ Sealing the surface of concrete
      - * Film-forming sealers and floor paint
      - * Seamless polymer/resin flooring
      - * Bund wall/containment lining
    - ◇ Waterproofing and damp proofing of concrete walls
    - ◇ Roof coating
    - ◇ Concrete bridge deck membranes
    - ◇ Sealing and waterproofing of masonry
    - ◇ Preserving machinery, equipment and structures
      - * Maintenance coatings/paints for metals, alloys and concrete
      - * Chemical resistant coatings
    - ◇ Wear resistance
      - * Hard anti-scratch coating on plastics and other materials e.g. of titanium nitride to reduce scratching and abrasion loss
      - * Barrier coatings on concrete, metals and alloys subject to erosion/abrasive attack
  - » Anti-corrosion
    - * Underbody sealant for cars
    - * Preserving equipment and structural steel from degradation
    - * Under thermal insulation and under protective fireproofing for structural steel

- ◇ Passive fire protection

- ◇ Antimicrobial surface

- ◇ Foul release and anti-fouling

- Magnetic properties such as for magnetic media like cassette tapes, floppy disks, and some mass transit tickets

- Electrical or electronic properties

  - » Conformal Antenna, e.g., metal coatings on plastic airframes

  - » Conductive coatings e.g. to manufacture some types of resistors

  - » Insulating coatings e.g. on magnet wires used in transformers

- Scent properties such as scratch and sniff stickers and labels

## Passivation (Chemistry)

Passivation, in physical chemistry and engineering, refers to a material becoming "passive," that is, less affected or corroded by the environment of future use. Passivation involves creation of an outer layer of shield material that is applied as a microcoating, created by chemical reaction with the base material, or allowed to build from spontaneous oxidation in the air. As a technique, passivation is the use of a light coat of a protective material, such as metal oxide, to create a shell against corrosion. Passivation can occur only in certain conditions, and is used in microelectronics to enhance silicon. The technique of passivation strengthens and preserves the appearance of metallics. In electrochemical treatment of water, passivation reduces the effectiveness of the treatment by increasing the circuit resistance, and active measures are typically used to overcome this effect, the most common being polarity reversal, which results in limited rejection of the fouling layer. Other proprietary systems to avoid electrode passivation, several discussed below, are the subject of ongoing research and development.

When exposed to air, many metals naturally form a hard, relatively inert surface, as in the tarnish of silver. In the case of other metals, such as iron, a somewhat rough porous coating is formed from loosely adherent corrosion products. In this case, a substantial amount of metal is removed, which is either deposited or dissolved in the environment. Corrosion coating reduces the rate of corrosion by varying degrees, depending on the kind of base metal and its environment, and is notably slower in room-temperature air for aluminium, chromium, zinc, titanium, and silicon (a metalloid); the shell of corrosion inhibits deeper corrosion, and operates as one form passivation. The inert surface layer, termed the "native oxide layer", is usually an oxide or a nitride, with a thickness of a monolayer (1-3 Å) for a noble metal such as platinum, about 15 Å for silicon, and nearer to 50 Å for aluminium after several years.

# Mechanisms

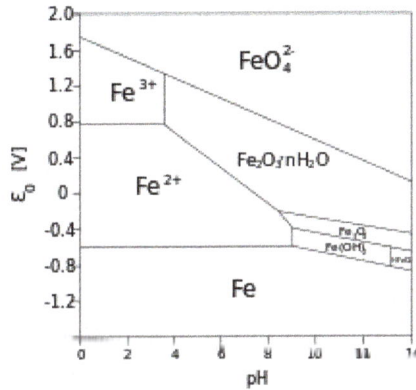

Pourbaix diagram of iron.

There has been much interest in determining the mechanisms that govern the increase of thickness of the oxide layer over time. Some of the important factors are the volume of oxide relative to the volume of the parent metal, the mechanism of oxygen diffusion through the metal oxide to the parent metal, and the relative chemical potential of the oxide. Boundaries between micro grains, if the oxide layer is crystalline, form an important pathway for oxygen to reach the unoxidized metal below. For this reason, vitreous oxide coatings – which lack grain boundaries – can retard oxidation. The conditions necessary (but not sufficient) for passivation are recorded in Pourbaix diagrams. Some corrosion inhibitors help the formation of a passivation layer on the surface of the metals to which they are applied. Some compounds, dissolving in solutions (chromates, molybdates) form non-reactive and low solubility films on metal surfaces.

# Discovery

In the mid 1800s, Christian Friedrich Schönbein discovered that when a piece of iron is placed in dilute nitric acid, it will dissolve and produce hydrogen, but if the iron is placed in concentrated nitric acid and then returned to the dilute nitric acid, little or no reaction will take place. Schönbein named the first state the active condition and the second the passive condition. If passive iron is touched by active iron, it becomes active again. In 1920, Ralph S. Lillie measured the effect of an active piece of iron touching a passive iron wire and found that "a wave of activation sweeps rapidly (at some hundred centimeters a second) over its whole length".

# Specific Materials

## Silicon

In the area of microelectronics, the formation of a strongly adhering passivating oxide is important to the performance of silicon.

In the area of photovoltaics, a passivating surface layer such as silicon nitride, silicon dioxide or titanium dioxide can reduce surface recombination - a significant loss mechanism in solar cells.

## Aluminium

Pure aluminium naturally forms a thin surface layer of aluminium oxide on contact with oxygen in the atmosphere through a process called oxidation, which creates a physical barrier to corrosion or further oxidation in most environments. Aluminium alloys, however, offer little protection against corrosion. There are three main ways to passivate these alloys: *alclading, chromate conversion coating* and *anodizing.* Alclading is the process of metallurgically bonding a thin layer of pure aluminium to the aluminium alloy. Chromate conversion coating is a common way of passivating not only aluminum, but also zinc, cadmium, copper, silver, magnesium, and tin alloys. Anodizing forms a thick oxide coating. This finish is more robust than the other processes and also provides good electrical insulation, which the other two processes do not.

For example, prior to storing hydrogen peroxide in an aluminium container, the container can be passivated by rinsing it with a dilute solution of nitric acid and peroxide alternating with deionized water. The nitric acid and peroxide oxidizes and dissolves any impurities on the inner surface of the container, and the deionized water rinses away the acid and oxidized impurities.

## Ferrous Materials

Ferrous materials, including steel, may be somewhat protected by promoting oxidation ("rust") and then converting the oxidation to a metalophosphate by using phosphoric acid and further protected by surface coating. As the uncoated surface is water-soluble, a preferred method is to form manganese or zinc compounds by a process commonly known as Parkerizing or phosphate conversion. Older, less-effective but chemically-similar electrochemical conversion coatings included black oxidizing, historically known as bluing or browning. Ordinary steel forms a passivating layer in alkali environments, as reinforcing bar does in concrete.

## Stainless Steel

Stainless steels are corrosion-resistant by nature, which might suggest that passivating them would be unnecessary. However, stainless steels are not completely impervious to rusting. One common mode of corrosion in corrosion-resistant steels is when small spots on the surface begin to rust because grain boundaries or embedded bits of foreign matter (such as grinding swarf) allow water molecules to oxidize some of the iron in those spots despite the alloying chromium. This is called rouging. Some grades of stainless steel are especially resistant to rouging; parts made from them may therefore forgo any passivation step, depending on engineering decisions.

Passivation processes are generally controlled by industry standards, the most prevalent among them today being ASTM A 967 and AMS 2700. These industry standards generally list several passivation processes that can be used, with the choice of specific method left to the customer and vendor. The "method" is either a nitric acid-based passivating bath, or a citric acid-based bath. The various 'types' listed under each method refer to differences in acid bath temperature and concentration. Sodium dichromate is often required as an additive to promote oxidation in certain 'types' of nitric-based acid baths.

Common among all of the different specifications and types are the following steps: Prior to passivation, the object must be cleaned of any contaminants and generally must undergo a validating test to prove that the surface is 'clean.' The object is then placed in an acidic passivating bath that meets the temperature and chemical requirements of the method and type specified between customer and vendor. (Temperatures can range from ambient to 60 degrees C {140 degrees F}), while minimum passivation times are usually 20 to 30 minutes). The parts are neutralized using a bath of aqueous sodium hydroxide, then rinsed with clean water and dried. The passive surface is validated using humidity, elevated temperature, a rusting agent (salt spray), or some combination of the three. However, proprietary passivation processes exist for martensitic stainless steel, which is difficult to passivate, as microscopic discontinuities can form in the surface of a machined part during passivation in a typical nitric acid bath. The passivation process removes exogenous iron, creates/restores a passive oxide layer that prevents further oxidation (rust), and cleans the parts of dirt, scale, or other welding-generated compounds (e.g. oxides).

It is not uncommon for some aerospace manufacturers to have additional guidelines and regulations when passivating their products that exceed the national standard. Often, these requirements will be cascaded down using Nadcap or some other accreditation system. Various testing methods are available to determine the passivation (or passive state) of stainless steel. The most common methods for validating the passivity of a part is some combination of high humidity and heat for a period of time, intended to induce rusting. Electro-chemical testers can also be utilized to commercially verify passivation.

## Nickel

Nickel can be used for handling elemental fluorine, owing to the formation of a passivation layer of nickel fluoride. This fact is useful in water treatment and sewage treatment applications.

# Low Plasticity Burnishing

Low plasticity burnishing (LPB) is a method of metal improvement that provides deep, stable surface compressive residual stresses with little cold work for improved damage tolerance and metal fatigue life extension. Improved fretting fatigue and stress corrosion performance has been documented, even at elevated temperatures where the compression from other metal improvement processes relax. The resulting deep layer of compressive residual stress has also been shown to improve high cycle fatigue (HCF) and low cycle fatigue (LCF) performance.

## History

Unlike LPB, traditional burnishing tools consist of a hard wheel or fixed lubricated ball pressed into the surface of an asymmetrical work piece with sufficient force to deform the surface layers, usually in a lathe. The process does multiple passes over the work pieces, usually under increasing load, to improve surface finish and deliberately cold work the surface. Roller and ball burnishing have been studied in Russia and Japan, and were applied most extensively in the USSR in the 1970s. Various burnishing methods are used, particularly in Eastern Europe, to improve fatigue life. Improvements in HCF, corrosion fatigue and SCC are documented, with fatigue strength enhancement attributed to improved finish, the development of a compressive surface layer, and the increased yield strength of the cold worked surface.

LPB was developed and patented by Lambda Technologies, a family-owned company from Cincinnati, Ohio, in 1996. Since then, LPB has been developed to produce compression in a wide array of materials to mitigate surface damage, including fretting, corrosion pitting, stress corrosion cracking (SCC), and foreign object damage (FOD), and is being employed to aid in daily MRO operations. To this day, LPB is the *only* metal improvement method applied under continuous closed-loop process control and has been successfully applied to turbine engines, piston engines, propellers, aging aircraft structures, landing gear, nuclear waste material containers, biomedical implants and welded joints. The applications involved titanium, iron, nickel and steel-based components and showed improved damage tolerance as well as high and low cycle fatigue performance by an order of magnitude.

## How it Works

The basic LPB tool is a ball that is supported in a spherical hydrostatic bearing. The tool can be held in any CNC machine or by industrial robots, depending on the application. The machine tool coolant is used to pressurize the bearing with a continuous flow of fluid to support the ball. The ball does not contact the mechanical bearing seat, even under load. The ball is loaded at a normal state to the surface of a component with a hydraulic cylinder that is in the body of the tool. LPB can be performed in conjunction with chip forming machining operations in the same CNC machining tool.

The ball rolls across the surface of a component in a pattern defined in the CNC code, as in any machining operation. The tool path and normal pressure applied are designed to create a distribution of compressive residual stress. The form of the distribution is designed to counter applied stresses and optimize fatigue and stress corrosion performance. Since there is no shear being applied to the ball, it is free to roll in any direction. As the ball rolls over the component, the pressure from the ball causes plastic deformation to occur in the surface of the material under the ball. Since the bulk of the material constrains the deformed area, the deformed zone is left in compression after the ball passes.

## Benefits

The LPB process includes a unique and patented way of analyzing, designing, and testing metallic components in order to develop the unique metal treatment necessary to improve performance and reduce metal fatigue, SCC, and corrosion fatigue failures. Lambda designs a new tool for each component to provide the best results possible and to ensure that the apparatus reaches every inch on the component. With this practice of constantly redesigning, along with the closed-loop process control system, LPB has been shown to produce a maximum compression of 12mm, although the average is around 1-7+mm. LPB has even been shown to have the ability to produce through-thickness compression in blades and vanes, greatly increasing their damage tolerance over 10-fold, effectively mitigating most FOD and reducing inspection requirements. No material is removed during this process, even when correcting corrosion damage. LPB smooths surface asperities during machining, leaving an improved, almost mirror-like surface finish that is vastly better looking and better protected than even a newly manufactured component.

## Cold Working

The cold work produced from this process is typically minimal, similar to the cold work produced by laser peening, only a few percent, but a great deal less than shot peening, gravity peening or, deep rolling. Cold work is particularly important because the higher the cold work at the surface of a component, the more vulnerable to elevated temperatures and mechanical overload that component will be and the easier the beneficial surface residual compression will relax, rendering the treatment pointless. In other words, a component that has been highly cold worked will not hold the compression if it comes into contact with extreme heat, like an engine, and will be just as vulnerable as it was to start. Therefore, LPB and laser peening stand out in the surface enhancement industry because they are both thermally stable at high temperatures. The reason LPB produces such low percentages of cold work is because of the aforementioned closed-loop process control. Conventional shot peening processes have some guesswork involved and are not exact at all, causing the procedure to have to be performed multiple times on one component. For example, shot peening, in order to make sure every spot on the component is treated, typically specifies coverage of between 200% (2T) and 400% (4T). This means that at 200% coverage (2T), 5 or more impacts occur at 84% of locations and at 400% coverage (4T), it is significantly more. The problem is that one area will be hit several times while the area next to it is hit fewer times, leaving uneven compression at the surface. This uneven compression results in the whole process being easily "undone", as was mentioned above. LPB requires only one pass with the tool and leaves a deep, even, beneficial compressive stress.

The LPB process can be performed on-site in the shop or in situ on aircraft using robots, making it easy to incorporate into everyday maintenance and manufacturing procedures. The method is applied under continuous closed loop process control (CLPC), creating accuracy within 0.1% and alerting the operator and QA immediately if the processing bounds are exceeded. The limitation of this process is that different CNC processing codes need to be developed for each application, just like any other machining task. The other issue is that because of dimensional restrictions, it may not be possible to create the tools necessary to work on certain geometries, although that has yet to be a problem.

## DCVG

DCVG stands for Direct Current Voltage Gradient and is a survey technique used for assessing the effectiveness of corrosion protection on buried steel structures. In particular, oil and natural gas pipelines are routinely monitored using this technique to help locate coating faults and highlight deficiencies in their cathodic protection (CP) strategies.

### History

The DCVG method was invented by Australian John Mulvany, an ex Telecom engineer, in the early 1980s. This technique was used by Telecom Australia to identify damaged insulation on buried metallic cable. At that time Santos in Adelaide was keen to utilise coating defect techniques for buried pipelines suffering corrosion in the Moomba area. Dr John Leeds, a professional corrosion engineer, was employed by Santos to engage companies with relevant expertise. Initially international companies utilising the "CIPS" and "Pearson" technique were engaged.

Ike Solomon and Matthew Wong of Wilson Walton International engaged John Mulvaney to modify the DCVG technique to make it applicable for buried pipelines. Field testing of the method was first performed on the Shell White Oil Pipeline. Subsequently, trials were performed for both Santos and The Pipeline Authority of South Australia. Vastly superior results were obtained over the other techniques. Ike Solomon and Bob Phang of Solomon Corrosion Consulting Services first demonstrated the technique overseas in the USA and Canada in 1985.

Today, the DCVG technique is universally accepted throughout the pipeline industry and is described in NACE International test method TM-0109-2009. Industry codes referring to pipe/pipeline inspection (such as API 571 and API RP 574, published by the American Petroleum Institute) reference it as a suitable method for determining coating breakdown in buried pipelines.

### Background

Buried steel structures will eventually corrode if not provided corrosion control and the rate of corrosion can be unacceptably rapid in some soils or where exposed to salt water. The primary form of corrosion protection is usually one or more protective coatings, such as epoxy, bitumen, resin etc. For buried pipelines (for example), coatings alone are insufficient as corrosion will likely occur at defects and corrosion control is commonly supplemented by cathodic protection. As pipelines age coatings deteriorate and the cathodic protection becomes increasingly important in mitigating corrosion damage. Prior to the use of DCVG, assessing the condition of the pipeline coating(s) was performed using indirect techniques like close interval potential surveys or expensive excavations of the pipeline.. The DCVG technique was developed to locate coating faults, quantify their severity and measure the effectiveness of the Cathodic protection used without having to disturb the pipeline.

## Principle

Assuming that the buried pipeline is protected using Impressed Current Cathodic Protection (ICCP - as are most pipelines with hazardous contents), then any defects in the coating will result in electric current flowing from the surrounding soil and into the pipe. These currents cause voltage gradients to be set up in the soil, which can be measured using a voltmeter. By looking at the direction of these gradients, the location of coating faults may be identified. By plotting the direction of voltage gradients around a fault, the type and nature of faults may be deduced. By measuring the localised soil potentials with respect to remote earth, a measure of the effectiveness of the cathodic protection may be calculated.

## Practical Methods

In theory, a standard analogue electronic multimeter could be used to perform a DCVG survey, but in practice it would be very difficult to take accurate readings and assess the direction of the voltage gradients correctly. A digital multimeter is completely unsuitable because of the difficulty in quickly assessing the direction of the voltage gradient. Specially designed DCVG meters are available, which have bespoke voltage ranges, specially designed transient response, rugged cases and (usually) a centre-zero meter movement for ease of use. The NACE method requires the measurements to be made using a pair of copper-copper(II) sulfate electrodes rather than simple metallic probes. In addition, the cathodic protection is switched on and off repeatedly using an electronic switch commonly referred to as an interrupter. Thus, two voltage readings (the "on" and "off" potentials) are taken at each fault position. Counter-intuitively, it is actually the "off" potential which is regarded as more indicative of the effectiveness of the CP applied to the pipeline.

Pipelines which do not have any form of CP may be surveyed by using a temporary DC supply and anode bed. Long pipelines frequently have more than one DC supply for their CP, requiring a number of synchronised interrupters to perform a survey. DCVG surveys are often combined with other techniques, such as close interval potential survey and soil resistivity as part of a comprehensive corrosion protection program.

A Purpose-Built DCVG Meter

Results of a DCVG survey often result in selecting locations to excavate pipelines, which can be costly in urban areas. Collection of data and interpretation may be performed by pipeline companies themselves or, more usually, by independent specialists.

## Corrosion Mapping by Ultrasonics

Corrosion mapping by ultrasonics is a nonintrusive (noninvasive) technique which maps material thickness using ultrasonic techniques. Variations in material thickness due to corrosion can be identified and graphically portrayed as an image. The technique is widely used in the oil and gas industries for the in-service detection and characterization of corrosion in pipes and vessels. The data is stored on a computer and may be color coded to show differences in thickness readings. Corrosion may be mapped using Zero degree ultrasonic probes, an Eddy current array and/or Time of flight detection methods. In the book Nondestructive Examination of Underwater Welded Structures by Victor S. Davey describes a "fully automated dual axis robotic scanner used for corrosion mapping normally using a single zero degree compression probe scanned in a raster pattern over the area of interest." He also goes on to explain that "typically a 4 mm by 4 mm raster" is used.

## Cyclic Corrosion Testing

Cyclic Corrosion Testing (CCT) has evolved in recent years, largely within the automotive industry, as a way of accelerating real-world corrosion failures, under laboratory controlled conditions. As the name implies, the test comprises different climates which are cycled automatically so the samples under test undergo the same sort of changing environment that would be encountered in the natural world. The intention being to bring about the type of failure that might occur naturally, but more quickly i.e. accelerated. By doing this manufacturers and suppliers can predict, more accurately, the service life expectancy of their products.

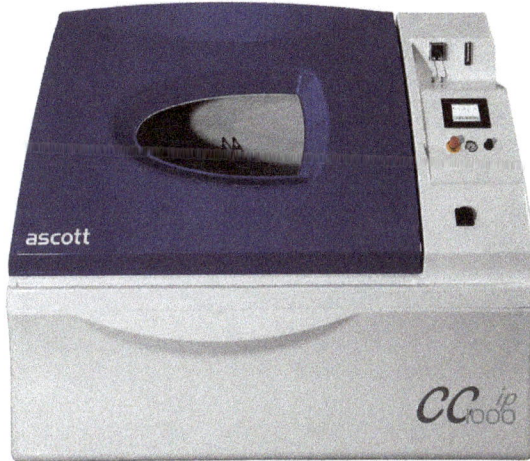

Example of a Cyclic corrosion test chamber.

Until the development of Cyclic Corrosion Testing, the traditional Salt spray test was virtually all that manufacturers could use for this purpose. However, this test was never intended for this purpose. Because the test conditions specified for salt spray testing are not typical of a naturally occurring environment, this type of test cannot be used as a reliable means of predicting the 'real world' service life expectancy for the samples under test. The sole purpose of the salt spray test is to compare and contrast results with previous experience to perform a quality audit. So, for example, a spray test can be used to 'police' a production process and forewarn of potential manufacturing problems or defects, which might affect corrosion resistance.

To recreate these different environments within an environmental chamber requires much more flexible testing procedures than are available in a standard salt spray chamber.

The lack of correlation between results obtained from traditional salt spray testing and the 'real world' atmospheric corrosion of vehicles, left the automotive industry without a reliable test method for predicting the service life expectancy of their products. This was and remains of particular concern in an industry where anti-corrosion warranties have been gradually increasing and now run to several years for new vehicles.

With ever increasing consumer pressure for improved vehicle corrosion resistance and a few 'high profile' corrosion failures amongst some vehicle manufactures – with disastrous commercial consequences, the automotive industry recognized the need for a different type of corrosion test.

Such a test would need to simulate the types of conditions a vehicle might encounter naturally, but recreate and accelerate these conditions, with good repeatability, within the convenience of the laboratory. CCT is effective for evaluating a variety of corrosion types, including galvanic corrosion and crevice corrosion.

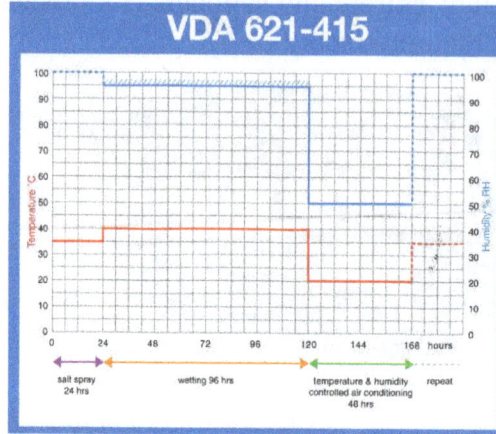

Graph showing the temperature & humidity steps required during cyclic corrosion test VDA 621-415

## Test Stages

Taking results gathered largely from 'real world' exposure sites, automotive companies, led originally by the Japanese automobile industry, developed their own Cyclic Corrosion Tests. These have evolved in different ways for different vehicle manufacturers, and such tests still remain largely industry specific, with no truly international CCT standard. However, they all generally require most of the following conditions to be created, in a repeating sequence or 'cycle', though not necessarily in the following order:

- A salt spray 'pollution' phase. This may be similar to the traditional salt spray test although in some cases direct impingement by the salt solution on the test specimens, or even complete immersion in salt water, is required. However, this 'pollution' phase is generally shorter in duration than a traditional salt spray test.

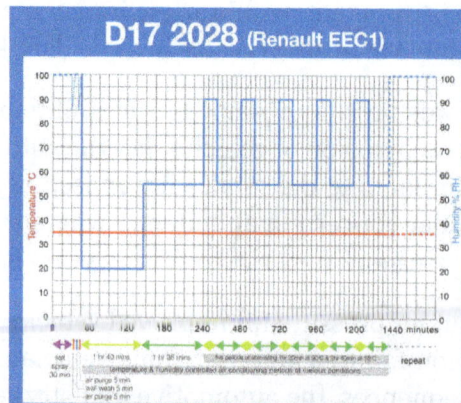

Graph showing the temperature & humidity steps required during cyclic corrosion test D17 2028 ECC1

- An air drying phase. Depending on the test, this may be conducted at ambient temperature, or at an elevated temperature, with or without control over the relative humidity and usually by introducing a continuous supply of relatively fresh air around the test samples at the same time. It is generally required that the samples under test should be visibly 'dry' at the end of this test phase.

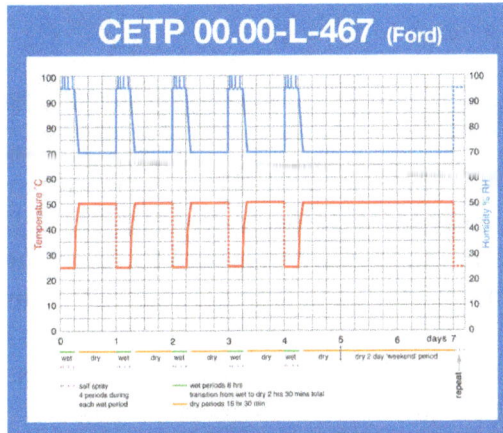

Graph showing the temperature & humidity steps required
during cyclic corrosion test CETP 00.00-L-467

- A condensation humidity 'wetting' phase. This is usually conducted at an elevated temperature and generally a high humidity of 95-100%RH. The purpose of this phase is to promote the formation of condensation on the surfaces of the samples under test.

- A controlled humidity/humidity cycling phase. This requires the tests samples to be exposed to a controlled temperature and controlled humidity climate, which can either be constant or cycling between different levels. When cycling between different levels, the rate of change may also be specified.

The above list is not exhaustive, since some automotive companies may also require other climates to be created in sequence as well, for example; sub-zero refrigeration, but it does list the most common requirements.

## Tests Standards

The below list is not exhaustive, but here are some examples of popular cyclic corrosion test standards:

- ACT 1 (Volvo)

- ACT 2 (Volvo)

- CETP 00.00-L-467 (Ford)

- D17 2028 (Renault)

- JASO M 609

- SAE J 2334

- VDA 621-415

# Pitting Resistance Equivalent Number

Pitting resistance equivalent number (PREN) is a predictive measurement of the corrosion resistance of various types of stainless steel. In general: the higher PREN-value, the more corrosion resistant the steel. Steels with PREN-values above 32 are considered seawater (corrosion) resistant. Furthermore a PREN-value ≥ 40 for duplex steels is called for in the DIN EN ISO 15156 as well as the American NACE – a standard for use in hydrogen sulfide environments known in the oil and gas extraction industries. Steels with PREN less than 32 are not sea water corrosion resistant. It depends on type:

- Ferritic stainless steels require a minimum PREN of 35

- Duplex stainless a minimum PREN of 40

- Super austenitic stainless a minimum PREN of 45

These alloys also need to be manufactured and heat treated correctly to be sea water corrosion resistant to the expected level. PREN alone is not an indicator of corrosion resistance. The value should be calculated for each heat to ensure compliance with minimum requirements, this is due to the chemistry variation within the same alloy compositional limits.

## PREN Formulas (w/w)

The PREN-value is calculated using the following formula: PREN = 1 x %Cr + 3.3 x %Mo + 16 x %N

Exception: stainless steels with molybdenum content ≥ 1.5% may have a PREN-value ≥ 30. In these norms the PREN-value takes into account tungsten [W] in the alloy and is defined with the formula: PREN = 1 · %Cr + 3.3 ( %Mo + 0.5 · %W ) + 16 · %N

*NOTE: There is another formula presented that takes Silicon [S] into account as well. Silicon presents at half the amount as molybdenum, meaning you need twice as much silicon to replace the same amount of molybdenum.*

This formula would be: PREN = 1 x %Cr + (3.3 x (%Mo + (1/2 x %Si) ) + 16 x %N

# Salt Spray Test

The salt spray (or salt fog) test is a standardized and popular corrosion test method, used to check corrosion resistance of materials and surface coatings. Usually, the materials to be tested are metallic (although stone, ceramics, and polymers may also be tested) and finished with a surface coating which is intended to provide a degree of corrosion protection to the underlying metal. Salt spray testing is an accelerated corrosion test that produces a corrosive attack to coated samples in order to evaluate (mostly comparatively) the suitability of the coating for use as a protective finish. The appearance of corrosion products (rust or other oxides) is evaluated after a pre-determined period of time. Test duration depends on the corrosion resistance of the coating; generally, the more corrosion resistant the coating is, the longer the period of testing before the appearance of corrosion/ rust. The salt spray test is one of the most widespread and long established corrosion tests. ASTM B117 was the first internationally recognized salt spray standard, originally published in 1939. Other important relevant standards are ISO9227, JIS Z 2371 and ASTM G85.

## Application

Salt spray testing is popular because it is relatively inexpensive, quick, well standardized, and reasonably repeatable. Although there may be a weak correlation between the duration in salt spray test and the expected life of a coating in certain coatings such as hot dip galvanized steel, this test has gained worldwide popularity due to low cost and quick results. Most Salt Spray Chambers today are being used NOT to predict the corrosion resistance of a coating, but to maintain coating processes such as pre-treatment and painting, electroplating, galvanizing, and the like, on a comparative basis. For example, pre-treated + painted components must pass 96 hours Neutral Salt Spray, to be accepted for production. Failure to meet this requirement implies instability in the chemical process of the pre-treatment, or the paint quality, which must be addressed immediately, so that the upcoming batches are of the desired quality. The longer the accelerated corrosion test, the longer the process remains out of control, and larger is the loss in the form of non-conforming batches. The principle application of the salt spray test is therefore enabling quick comparisons to be made between actual and expected corrosion resistance. Most commonly, the time taken for oxides to appear on the samples under test is compared to expectations, to determine whether the test is passed or failed. For this reason the salt spray test is most often deployed in a quality audit role, where, for example, it can be used to check the effectiveness of a production process, such as the surface coating of a metallic part. The salt spray test has little application in predicting how materials or surface coatings will resist corrosion in the real-world, because it does not create, replicate or accelerate real-world corrosive conditions. Cyclic corrosion testing is better suited to this.

# Testing Equipment

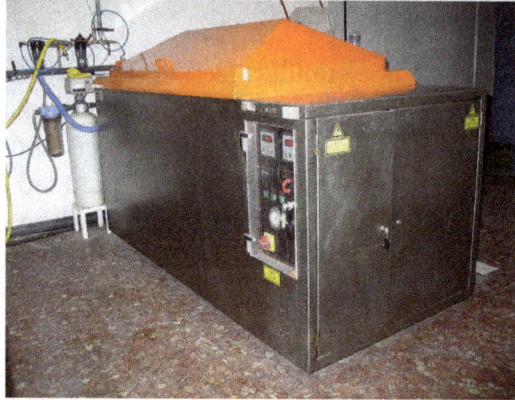

A salt spray cabinet

The apparatus for testing consists of a closed testing cabinet/chamber, where a salt water (5% NaCl) solution is atomized by means of spray nozzle(s) using pressurized air. This produces a corrosive environment of dense salt water fog (also referred to as a mist or spray) in the chamber, so that test samples exposed to this environment are subjected to severely corrosive conditions. Chamber volumes vary from supplier to supplier. If there is a minimum volume required by a particular salt spray test standard, this will be clearly stated and should be complied with. There is a general historical consensus that larger chambers can provide a more homogeneous testing environment.

Variations to the salt spray test solutions depend upon the materials to be tested. The most common test for steel based materials is the Neutral Salt Spray test (often abbreviated to NSS) which reflects the fact that this type of test solution is prepared to a neutral pH of 6.5 to 7.2. Results are represented generally as testing hours in NSS without appearance of corrosion products (e.g. 720 h in NSS according to ISO 9227). Synthetic seawater solutions are also commonly specified by some companies and standards. Other test solutions have other chemicals added including acetic acid (often abbreviated to ASS) and acetic acid with copper chloride (often abbreviated to CASS) each one chosen for the evaluation of decorative coatings, such as electroplated copper-nickel-chromium, electroplated copper-nickel or anodized aluminum. These acidified test solutions generally have a pH of 3.1 to 3.3.

Some sources do not recommend using ASS or CASS test cabinets interchangeably for NSS tests, due to the risk of cross-contamination, it is claimed that a thorough cleaning of the cabinet after CASS test is very difficult. ASTM does not address this issue, but ISO 9227 does not recommend it and if it is to be done, advocates a thorough cleaning.

Although the majority of salt spray tests are continuous, i.e.; the samples under test are exposed to the continuous generation of salt fog for the entire duration of the test, a few do not require such exposure. Such tests are commonly referred to as modified salt spray tests. ASTM G85 is an example of a test standard which contains several modified salt spray tests which are variations to the basic salt spray test.

## Modified Salt Spray Tests

ASTM G85 is the most popular global test standard covering modified salt spray tests. There are five such tests altogether, referred to in ASTM G85 as annexes A1 through to A5.

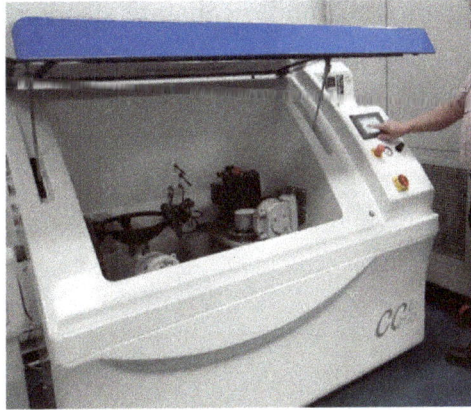

A modified salt spray chamber in use

Many of these modified tests originally arose within particular industry sector, in order to address the need for a corrosion test capable of replicating the effects of naturally occurring corrosion and accelerate these effects.

This acceleration arises through the use of chemically altered salt spray solutions, often combined with other test climates and in most cases, the relatively rapid cycling of these test climates over time. Although popular in certain industries, modified salt spray testing has in many cases been superseded by Cyclic corrosion testing (CCT) The type of environmental test chambers used for modified salt spray testing to ASTM G85 are generally similar to the chambers used for testing to ASTM B117, but will often have some additional features, such as an automatic climate cycling control system.

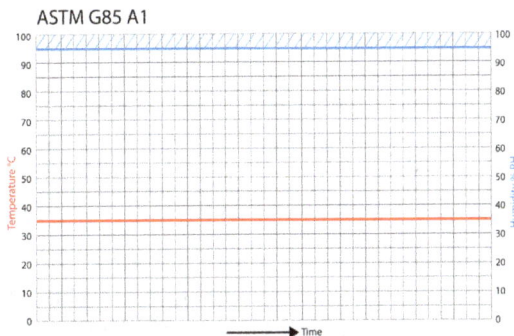

Graph showing the temperature & humidity steps required during
modified Salt Spray Test ASTM G85 Annex 1

ASTM G85 annex A1 – Acetic Acid Salt Spray Test (non-cyclic) This test can be used to determine the relative resistance to corrosion of decorative chromium plating on steel and zinc based die casting when exposed to an acetic acid salt spray climate at an elevated temperature. This test is also referred to as an ASS test. Test specimens are placed in an enclosed chamber and exposed to a continuous indirect spray of salt water solution, prepared in accordance with the requirements of the test standard and acidified (to pH 3.1 to 3.3) by the addition of acetic acid. This spray is set to fall-out on to the specimens at a rate of 1.0 to 2.0ml/80 cm²/hour, in a chamber temperature of +35C. This climate is maintained under constant steady state conditions. The test duration is variable.

Graph showing the temperature & humidity steps required during modified Salt Spray Test ASTM G85 Annex 2

ASTM G85 annex A2 – Acidified Salt Fog Test (cyclic).

This test can be used to test the relative resistance to corrosion of aluminium alloys when exposed to a changing climate of acetic acid salt spray, followed by air drying, followed by high humidity, all at an elevated temperature. This test is also referred to as a MASTMAASIS test. Test specimens are placed in an enclosed chamber, and exposed to a changing climate that comprises the following 3 part repeating cycle. 0.75 hours exposure to a continuous indirect spray of salt water solution, prepared in accordance with the requirements of the test standard and acidified (to pH 2.8 to 3.0) by the addition of acetic acid. This spray is set to fall-out on to the specimens at a rate of 1.0 to 2.0ml/80 cm²/hour. This is followed by 2.0 hours exposure to an air drying (purge) climate. This is followed by 3.25 hours exposure to a high humidity climate which gradually rises to between 65%RH and 95%RH. The entire test cycle is at a constant chamber temperature of +49C. The number of cycle repeats and therefore the test duration is variable.

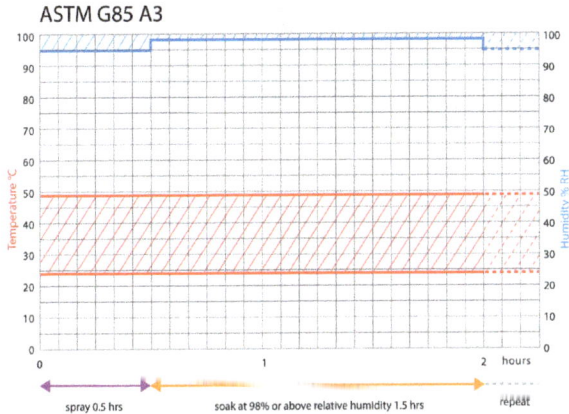

Graph showing the temperature & humidity steps required
during modified Salt Spray Test ASTM G85 Annex 3

## ASTM G85 annex A3 – Seawater Acidified Test (cyclic)

This test can be used to test the relative resistance to corrosion of coated or un-coated aluminium alloys and other metals, when exposed to a changing climate of acidified synthetic seawater spray, followed by a high humidity, both at an elevated temperature. This test is also referred to as a SWAAT test. Test specimens are placed in an enclosed chamber, and exposed to a changing climate that comprises the following 2 part repeating cycle. 30 minutes exposure to a continuous indirect spray of synthetic seawater solution, prepared in accordance with the requirements of the test standard and acidified (to pH 2.8 to 3.0) by the addition of acetic acid. This spray is set to fall-out on to the specimens at a rate of 1.0 to 2.0ml/80 cm²/hour. This is followed by 90 minutes exposure to a high humidity climate of above 98%RH. The entire test cycle is at a constant chamber temperature of +49C (may be reduced to +24 to +35C for organically coated specimens). The number of cycle repeats and therefore the test duration is variable.

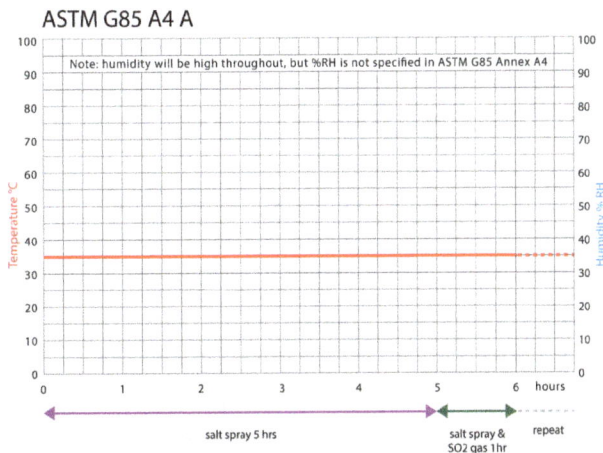

Graph showing the temperature & humidity steps required
during modified Salt Spray Test ASTM G85 Annex 4A

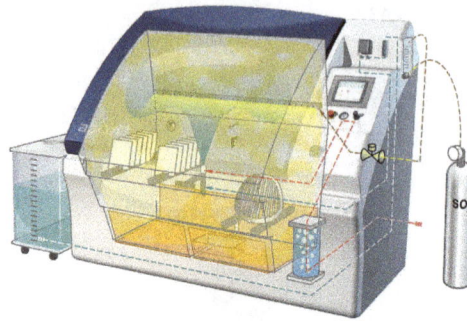

An example of modified salt spray test ASTM G85 annex A4 – SO2 Salt Spray Test

## ASTM G85 annex A4 – SO2 Salt Spray Test (cyclic)

This test can be used to test the relative resistance to corrosion of product samples that are likely to encounter a combined SO2(sulfur dioxide)/salt spray/acid rain environment during their usual service life. Test specimens are placed in an enclosed chamber, and exposed to 1 of 2 possible changing climate cycles. In either case, the exposure to salt spray may be salt water spray or synthetic sea water prepared in accordance with the requirements of the test standard. The most appropriate test cycle and spray solutions are to be agreed between parties.

The first climate cycle comprises a continuous indirect spray of neutral (pH 6.5 to 7.2) salt water/synthetic seawater solution, which falls-out on to the specimens at a rate of 1.0 to 2.0ml/80 cm²/hour. During this spraying, the chamber is dosed with SO2 gas at a rate of 35 cm³/minute/m³ of chamber volume, for 1 hour in every 6 hours of spraying. The entire test cycle is at a constant chamber temperature of +35C. The number of cycle repeats and therefore the test duration is variable.

The second climate cycle comprises 0.5 hours of continuous indirect spray of neutral (pH 6.5 to 7.2) salt water/synthetic seawater solution, which falls-out on to the specimens at a rate of 1.0 to 2.0ml/80 cm²/hour. This is followed by 0.5 hours of dosing with SO2 gas at a rate of 35 cm³/minute/m³ of chamber volume. This is followed by 2.0 hours of high humidity soak. The entire test cycle is at a constant chamber temperature of +35C. The number of cycle repeats and therefore the test duration is variable.

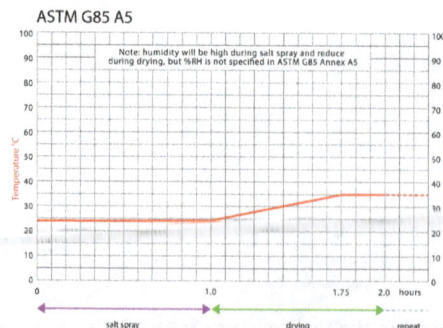

Graph showing the temperature & humidity steps required
during modified Salt Spray Test ASTM G85 Annex 5

ASTM G85 annex A5 - Dilute Electrolyte Salt Fog/Dry Test (cyclic)

This test can be used to test the relative resistance to corrosion paints on steel, when exposed to a changing climate of dilute salt spray at ambient temperature, followed by air drying at and elevated temperature. It is a popular test in the surface coatings industry, where it is also referred to as the PROHESION™ test. Test specimens are placed in an enclosed chamber, and exposed to a changing climate that comprises the following 2 part repeating cycle. 1.0 hour exposure to a continuous indirect spray of salt water solution, prepared in accordance with the requirements of the test standard and acidified (to pH 3.1 to 3.3) by the addition of acetic acid. This spray is set to fall-out on to the specimens at a rate of 1.0 to 2.0ml/80 cm²/hour, in an ambient chamber temperature (21 to 27C). This is followed by 1.0 hour exposure to an air drying (purge) climate, in a chamber temperature of +35C. The number of cycle repeats and therefore the test duration is variable.

## Standardization

Electroplated and yellow chromated bolt with white corrosion

Zinc flake coated bolt with red rust after testing

Chamber construction, testing procedure and testing parameters are standardized under national and international standards, such as ASTM B 117 and ISO 9227. These standards describe the necessary information to carry out this test; testing parameters such as temperature, air pressure of the sprayed solution, preparation of the spraying solution, concentration, pH, etc. Daily checking of testing parameters is necessary to show compliance with the standards, so records shall be maintained accordingly. ASTM B117 and ISO 9227 are widely used as reference standards. Testing cabinets are manufactured according to the specified requirements here.

However, these testing standards neither provide information of testing periods for the coatings to be evaluated, nor the appearance of corrosion products in form of salts. Requirements are agreed between customer and manufacturer. In the automotive industry requirements are specified under material specifications. Different coatings have different behavior in salt spray test and consequently, test duration will differ from one type of coating to another. For example, a typical electroplated zinc and yellow passivated steel part lasts 96 hours in salt spray test without white rust. Electroplated zinc-nickel steel parts can last more than 720 hours in NSS test without red rust (or 48 hours in CASS test without red rust) Requirements are established in test duration (hours) and coatings shall comply with minimum testing periods.

Artificial seawater which is sometimes used for Salt Spray Testing can be found at ASTM International. The standard for Artificial Seawater is ASTM D1141-98 which is the standard practice for the preparation of substitute ocean water.

## Uses

Typical coatings that can be evaluated with this method are:

- Phosphated (pre-treated) surfaces (with subsequent paint/primer/lacquer/rust preventive)

- Zinc and zinc-alloy plating.

- Electroplated chromium, nickel, copper, tin

- Coatings not applied electrolytically, such as zinc flake coatings according to ISO 10683

- Organic coatings, such as rust preventives

- Paint Coating

Hot-dip galvanized surfaces are not generally tested in a salt spray test. Hot-dip galvanizing produces zinc carbonates when exposed to a natural environment, thus protecting the coating metal and reducing the corrosion rate. The zinc carbonates are not produced when a hot-dip galvanized specimen is exposed to a salt spray fog, therefore this testing method does not give an accurate measurement of corrosion protection. ISO 9223 gives the guidelines for proper measurement of corrosion resistance for hot-dip galvanized specimens.

Painted surfaces with an underlying hot-dip galvanized coating can be tested according to this method.

Testing periods range from a few hours (e.g. 8 or 24 hours of phosphated steel) to more than a month (e.g. 720 hours of zinc-nickel coatings, 1000 hours of certain zinc flake coatings).

# References

- Lillie, Ralph S. (June 20, 1920). "The Recovery of Transmissivity in Passive Iron Wires as a Model of Recovery Processes in Irritable Living Systems". The Journal of General Physiology. Physiological Laboratory, Clark University, Worcester. 3 (2): 129–43. doi:10.1085/jgp.3.2.129. Retrieved 15 August 2015

- S. Grainger and J. Blunt, Engineering Coatings: Design and Application, Woodhead Publishing Ltd, UK, 2nd ed., 1998, ISBN 978-1-85573-369-5

- "The Recycling of Stainless Steel ("Recycled Content" and "Input Composition" slides)" (Flash). International Stainless Steel Forum. 2006. Retrieved 19 November 2006

- Wu, Wenjie; Maye, Mathew M. (2014-01-01). "Void Coalescence in Core/Alloy Nanoparticles with Stainless Interfaces". Small. 10 (2): 271–276. doi:10.1002/smll.201301420

- Charalampos, Vasilatos, [et al.] (2008). "Hexavalent chromium and other toxic elements in natural waters in the Thiva – Tanagra – Malakasa Basin, Greece" (PDF). Hellenic Journal of Geosciences. 43 (57–56)

- Ashby, Michael F.; David R. H. Jones (1992) [1986]. "Chapter 12". Engineering Materials 2 (with corrections ed.). Oxford: Pergamon Press. p. 119. ISBN 0-08-032532-7

- "Room Temp Blackening Processes for Stainless Steel – Black Oxide – EPi". Electrochemical Products, Inc. Retrieved 2016-11-15

- Anusavice, Kenneth J. (2003) Phillips' Science of Dental Materials, 11th Edition. W.B. Saunders Company. ISBN 0721693873. p. 639

- "Facts About Chromium" (PDF). United States Environmental Protection Agency. April 13, 2013. Archived from the original on April 13, 2013. Retrieved July 13, 2016. CS1 maint: BOT: original-url status unknown (link)

- Hubert Gräfen, Elmar-Manfred Horn, Hartmut Schlecker, Helmut Schindler "Corrosion" Ullmann's Encyclopedia of Industrial Chemistry, Wiley-VCH: Weinheim, 2002. doi:10.1002/14356007.b01_08

- Hena, Sufia (2010-09-15). "Removal of chromium hexavalent ion from aqueous solutions using biopolymer chitosan coated with poly 3-methyl thiophene polymer". Journal of Hazardous Materials. 181 (1–3): 474–479. doi:10.1016/j.jhazmat.2010.05.037

- GalvInfo (August 2011). "GalvInfoNote / The Spangle on Hot-Dip Galvanized Steel Sheet" (PDF). GalvInfo. Retrieved 27 February 2014

- Edwards, Joseph (1997). Coating and Surface Treatment Systems for Metals. Finishing Publications Ltd. and ASM International. pp. 214–217. ISBN 0-904477-16-9

- "Dozens Of National Guard Soldiers Sick After Iraq 2003 Deploy, Toxic Chemical Eyed". Fox News. 2009-06-27. Retrieved 2016-10-27

- David Blowes (2002). "Tracking Hexavalent Cr in Groundwater". Science. 295 (5562): 2024–25. PMID 11896259. doi:10.1126/science.1070031

# Plating: An Integrated Study

Plating is a type of surface covering which has been practiced hundreds of years. It is used to improve solderability and to prevent corrosion. Electroplating, chrome plating, etc. are some topics explained in relation to plating. This chapter will provide an integrated understanding of plating.

## Plating

Plating is a surface covering in which a metal is deposited on a conductive surface. Plating has been done for hundreds of years; it is also critical for modern technology. Plating is used to decorate objects, for corrosion inhibition, to improve solderability, to harden, to improve wearability, to reduce friction, to improve paint adhesion, to alter conductivity, to improve IR reflectivity, for radiation shielding, and for other purposes. Jewelry typically uses plating to give a silver or gold finish. Thin-film deposition has plated objects as small as an atom, therefore plating finds uses in nanotechnology.

There are several plating methods, and many variations. In one method, a solid surface is covered with a metal sheet, and then heat and pressure are applied to fuse them (a version of this is Sheffield plate). Other plating techniques include electroplating, vapor deposition under vacuum and sputter deposition. Recently, plating often refers to using liquids. Metallizing refers to coating metal on non-metallic objects.

### Electroplating

In electroplating, an ionic metal is supplied with electrons to form a non-ionic coating on a substrate. A common system involves a chemical solution with the ionic form of the metal, an anode (positively charged) which may consist of the metal being plated (a soluble anode) or an insoluble anode (usually carbon, platinum, titanium, lead, or steel), and finally, a cathode (negatively charged) where electrons are supplied to produce a film of non-ionic metal.

### Electroless Plating

Electroless plating, also known as chemical or auto-catalytic plating, is a non-galvanic plating method that involves several simultaneous reactions in an aqueous solution, which occur without the use of external electrical power. The reaction is accomplished when hydrogen is released by a reducing agent, normally sodium hypophosphite (Note: the hydrogen leaves as a hydride ion) or thiourea, and oxidized, thus producing a negative charge on the surface of the part. The most common electroless plating method is electroless nickel plating, although silver, gold and copper layers can also be applied in this manner, as in the technique of Angel gilding.

## Specific Cases

## Gold Plating

Gold plating is a method of depositing a thin layer of gold on the surface of glass or metal, most often copper or silver.

Gold plating is often used in electronics, to provide a corrosion-resistant electrically conductive layer on copper, typically in electrical connectors and printed circuit boards. With direct gold-on-copper plating, the copper atoms have the tendency to diffuse through the gold layer, causing tarnishing of its surface and formation of an oxide/sulfide layer. Therefore, a layer of a suitable barrier metal, usually nickel, has to be deposited on the copper substrate, forming a copper-nickel-gold sandwich.

Metals and glass may also be coated with gold for ornamental purposes, using a number of different processes usually referred to as *gilding*.

Sapphires, plastics, and carbon fiber are some other materials that are able to be plated using advance plating techniques. The substrates that can be used are almost limitless.

## Silver Plating

A silver-plated alto saxophone

Silver plating has been used since the 18th century to provide cheaper versions of household items that would otherwise be made of solid silver, including cutlery, vessels of various kinds, and candlesticks. In the UK the assay offices, and silver dealers and collectors, use the term "silver plate" for items made from solid silver, derived long before silver plating was invented from the Spanish word for silver "plata", seizures of silver from Spanish ships carrying silver from America being a large source of silver at the time. This can cause confusion when talking about silver items; plate or plated. In the UK it is illegal to describe silver-plated items as "silver". It is not illegal to describe silver-plated items as "silver plate", although this is grammatically incorrect, and should also be avoided to prevent confusion.

The earliest form of silver plating was Sheffield Plate, where thin sheets of silver are fused to a layer or core of base metal, but in the 19th century new methods of production (including electroplating) were introduced. Britannia metal is an alloy of tin, antimony and copper developed as a base metal for plating with silver.

Another method that can be used to apply a thin layer of silver to objects such as glass, is to place Tollens' reagent in a glass, add Glucose/Dextrose, and shake the bottle to promote the reaction.

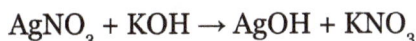

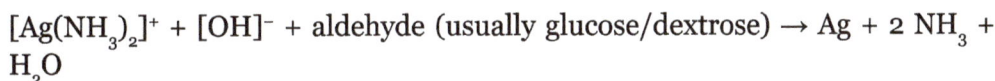

$$AgNO_3 + KOH \rightarrow AgOH + KNO_3$$

$$AgOH + 2\,NH_3 \rightarrow [Ag(NH_3)_2]^+ + [OH]^-$$

$$[Ag(NH_3)_2]^+ + [OH]^- + \text{aldehyde (usually glucose/dextrose)} \rightarrow Ag + 2\,NH_3 + H_2O$$

For applications in electronics, silver is sometimes used for plating copper, as its electrical resistance is lower. more so at higher frequencies due to the skin effect. Variable capacitors are considered of the highest quality when they have silver-plated plates. Similarly, silver-plated, or even solid silver cables, are prized in audiophile applications; however some experts consider that in practice the plating is often poorly implemented, making the result inferior to similarly priced copper cables.

Care should be used for parts exposed to high humidity environments. When the silver layer is porous or contains cracks, the underlying copper undergoes rapid galvanic corrosion, flaking off the plating and exposing the copper itself; a process known as red plague.

## Rhodium Plating

Rhodium plating is occasionally used on white gold, silver or copper and its alloys. A barrier layer of nickel is usually deposited on silver first, though in this case it is not to prevent migration of silver through rhodium, but to prevent contamination of the rhodium bath with silver and copper, which slightly dissolve in the sulfuric acid usually present in the bath composition.

## Chrome Plating

Chrome plating is a finishing treatment using the electrolytic deposition of chromium. The most common form of chrome plating is the thin, decorative *bright chrome*, which is typically a 10-μm layer over an underlying nickel plate. When plating on iron or steel, an underlying plating of copper allows the nickel to adhere. The pores (tiny holes) in the nickel and chromium layers work to alleviate stress caused by thermal expansion mismatch but also hurt the corrosion resistance of the coating. Corrosion resistance relies on what is called the passivation layer, which is determined by the chemical composition and processing, and is damaged by cracks and pores. In a special case, micropores can help distribute the electrochemical potential that accelerates galvanic corrosion between the layers of nickel and chromium. Depending on the application, coatings of different thicknesses will require different balances of the aforementioned properties. Thin, bright chrome imparts a mirror-like finish to items such as metal furniture frames and automotive trim. Thicker deposits, up to 1000 μm, are called *hard chrome* and are used in industrial equipment to reduce friction and wear.

The traditional solution used for industrial hard chrome plating is made up of about 250 g/L of $CrO_3$ and about 2.5 g/L of $SO_4^-$. In solution, the chrome exists as chromic acid, known as hexavalent chromium. A high current is used, in part to stabilize a thin layer of chromium(+2) at the surface of the plated work. Acid chrome has poor throwing power, fine details or holes are further away and receive less current resulting in poor plating.

## Zinc Plating

Zinc coatings prevent oxidation of the protected metal by forming a barrier and by acting as a sacrificial anode if this barrier is damaged. Zinc oxide is a fine white dust that (in contrast to iron oxide) does not cause a breakdown of the substrate's surface integrity as it is formed. Indeed, the zinc oxide, if undisturbed, can act as a barrier to further oxidation, in a way similar to the protection afforded to aluminum and stainless steels by their oxide layers. The majority of hardware parts are zinc-plated, rather than cadmium-plated.

## Zinc-Nickel Plating

Zinc-Nickel plating is one of the best corrosion resistant finishes available offering over 5 times the protection of conventional zinc plating and up to 1,500 hours of Neutral Salt Spray test performance. This plating is a combination of a High Nickel Zinc-Nickel alloy (10% - 15% nickel) and some variation of chromate. The most common mixed chromates include Hexavalent Iridescent, Trivalent or Black Trivalent Chromate. Used to protect steel, cast iron, brass, copper, and other materials, this acidic plating is an environmentally safe option. Hexavalent Chromate has been classified as a human carcinogen by the EPA and OSHA.

## Tin Plating

The tin-plating process is used extensively to protect both ferrous and nonferrous surfaces. Tin is a useful metal for the food processing industry since it is non-toxic, ductile and corrosion resistant. The excellent ductility of tin allows a tin coated base metal sheet to be formed into a variety of shapes without damage to the surface tin layer. It provides sacrificial protection for copper, nickel and other non-ferrous metals, but not for steel.

Tin is also widely used in the electronics industry because of its ability to protect the base metal from oxidation thus preserving its solderability. In electronic applications, 3% to 7% lead may be added to improve solderability and to prevent the growth of metallic "whiskers" in compression stressed deposits, which would otherwise cause electrical shorting. However, RoHS (Restriction of Hazardous Substances) regulations enacted beginning in 2006 require that no lead be added intentionally and that the maximum percentage not exceed 1%. Some exemptions have been issued to RoHS requirements in critical electronics applications due to failures which are known to have occurred as a result of tin whisker formation.

## Alloy Plating

In some cases, it is desirable to co-deposit two or more metals resulting in an electroplated alloy deposit. Depending on the alloy system, an electroplated alloy may be solid solution strengthened or precipitation hardened by heat treatment to improve the plating's physical and chemical properties. Nickel-Cobalt is a common electroplated alloy.

## Composite Plating

Metal matrix composite plating can be manufactured when a substrate is plated in a bath containing a suspension of ceramic particles. Careful selection of the size and composition of the particles can fine-tune the deposit for wear resistance, high temperature performance, or mechanical strength. Tungsten carbide, silicon carbide, chromium carbide, and aluminum oxide (alumina) are commonly used in composite electroplating.

## Cadmium Plating

Cadmium plating is under scrutiny because of the environmental toxicity of the cadmium metal. Cadmium plating is widely used in some applications in the aerospace, military, and aviation fields. However, it is being phased out due to its toxicity.

Cadmium plating (or "cad plating") offers a long list of technical advantages such as excellent corrosion resistance even at relatively low thickness and in salt atmospheres, softness and malleability, freedom from sticky and/or bulky corrosion products, galvanic compatibility with aluminum, freedom from stick-slip thus allowing reliable torquing of plated threads, can be dyed to many colors and clear, has good lubricity and solderability, and works well either as a final finish or as a paint base.

If environmental concerns matter, in most aspects cadmium plating can be directly replaced with gold plating as it shares most of the material properties, but gold is more expensive and cannot serve as a paint base.

## Nickel Plating

The chemical reaction for nickel plating is:

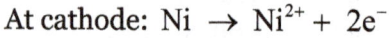

At cathode: $Ni \rightarrow Ni^{2+} + 2e^-$

At anode: $H_2PO_2 + H_2O \rightarrow H_2PO_3 + 2\ H^+$

Compared to cadmium plating, nickel plating offers a shinier and harder finish, but lower corrosion resistance, lubricity, and malleability, resulting in a tendency to crack or flake if the piece is further processed.

## Electroless Nickel Plating

Electroless nickel plating, also known as *enickel* and *NiP*, offers many advantages: uniform layer thickness over most complicated surfaces, direct plating of ferrous metals (steel), superior wear and corrosion resistance to electroplated nickel or chrome. Much of the chrome plating done in aerospace industry can be replaced with electroless nickel plating, again environmental costs, costs of hexavalent chromium waste disposal and notorious tendency of uneven current distribution favor electroless nickel plating.

Electroless nickel plating is self-catalyzing process, the resultant nickel layer is NiP compound, with 7–11% phosphorus content. Properties of the resultant layer hardness and wear resistance are greatly altered with bath composition and deposition temperature, which should be regulated with 1 °C precision, typically at 91 °C.

During bath circulation, any particles in it will become also nickel-plated; this effect is used to advantage in processes which deposit plating with particles like silicon carbide (SiC) or polytetrafluoroethylene (PTFE). While superior compared to many other plating processes, it is expensive because the process is complex. Moreover, the process is lengthy even for thin layers. When only corrosion resistance or surface treatment is of concern, very strict bath composition and temperature control is not required and the process is used for plating many tons in one bath at once.

Electroless nickel plating layers are known to provide extreme surface adhesion when plated properly. Electroless nickel plating is non-magnetic and amorphous. Electroless nickel plating layers are not easily solderable, nor do they seize with other metals or another electroless nickel-plated workpiece under pressure. This effect benefits electroless nickel-plated screws made out of malleable materials like titanium. Electrical resistance is higher compared to pure metal plating.

# Electroplating

Copper electroplating machine for layering PCBs

Electroplating is a process that uses electric current to reduce dissolved metal cations so that they form a thin coherent metal coating on an electrode. The term is also used for electrical oxidation of anions onto a solid substrate, as in the formation silver chloride on silver wire to make silver/silver-chloride electrodes. Electroplating is primarily used to change the surface properties of an object (e.g. abrasion and wear resistance, corrosion protection, lubricity, aesthetic qualities, etc.), but may also be used to build up thickness on undersized parts or to form objects by electroforming.

The process used in electroplating is called electrodeposition. It is analogous to a galvanic cell acting in reverse. The part to be plated is the cathode of the circuit. In one technique, the anode is made of the metal to be plated on the part. Both components are immersed in a solution called an electrolyte containing one or more dissolved metal salts as well as other ions that permit the flow of electricity. A power supply supplies a direct current to the anode, oxidizing the metal atoms that it comprises and allowing them to dissolve in the solution. At the cathode, the dissolved metal ions in the electrolyte solution are reduced at the interface between the solution and the cathode, such that they "plate out" onto the cathode. The rate at which the anode is dissolved is equal to the rate at which the cathode is plated, vis-a-vis the current through the circuit. In this manner, the ions in the electrolyte bath are continuously replenished by the anode.

Other electroplating processes may use a non-consumable anode such as lead or carbon. In these techniques, ions of the metal to be plated must be periodically replenished in the bath as they are drawn out of the solution. The most common form of electroplating is used for creating coins such as pennies, which are small zinc plates covered in a layer of copper.

## Process

The cations associate with the anions in the solution. These cations are reduced at the cathode to deposit in the metallic, zero valence state. For example, for copper plating, in an acid solution, copper is oxidized at the anode to $Cu^{2+}$ by losing two electrons. The $Cu^{2+}$ associates with the anion $SO_4^{2-}$ in the solution to form copper sulfate. At the cathode, the $Cu^{2+}$ is reduced to metallic copper by gaining two electrons. The result is the effective transfer of copper from the anode source to a plate covering the cathode.

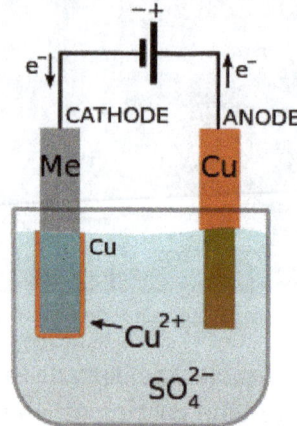

Electroplating of a metal (Me) with copper in a copper sulfate bath

The plating is most commonly a single metallic element, not an alloy. However, some alloys can be electrodeposited, notably brass and solder.

Many plating baths include cyanides of other metals (e.g., potassium cyanide) in addition to cyanides of the metal to be deposited. These free cyanides facilitate anode corrosion, help to maintain a constant metal ion level and contribute to conductivity. Additionally, non-metal chemicals such as carbonates and phosphates may be added to increase conductivity.

When plating is not desired on certain areas of the substrate, stop-offs are applied to prevent the bath from coming in contact with the substrate. Typical stop-offs include tape, foil, lacquers, and waxes.

The ability of a plating to cover uniformly is called *throwing power*; the better the "throwing power" the more uniform the coating.

## Strike

Initially, a special plating deposit called a "strike" or "flash" may be used to form a very thin (typically less than 0.1 micrometer thick) plating with high quality and good adherence to the substrate. This serves as a foundation for subsequent plating processes. A strike uses a high current density and a bath with a low ion concentration. The process is slow, so more efficient plating processes are used once the desired strike thickness is obtained.

The striking method is also used in combination with the plating of different metals. If it is desirable to plate one type of deposit onto a metal to improve corrosion resistance but this metal has inherently poor adhesion to the substrate, a strike can be first deposited that is compatible with both. One example of this situation is the poor adhesion of electrolytic nickel on zinc alloys, in which case a copper strike is used, which has good adherence to both.

## Electrochemical Deposition

Electrochemical deposition is generally used for the growth of metals and conducting metal oxides because of the following advantages: (i) the thickness and morphology of the nanostructure can be precisely controlled by adjusting the electrochemical parameters, (ii) relatively uniform and compact deposits can be synthesized in template-based structures, (iii) higher deposition rates are obtained, and (iv) the equipment is inexpensive due to the non-requirements of either a high vacuum or a high reaction temperature.

## Pulse Electroplating or Pulse Electrodeposition (PED)

A simple modification in the electroplating process is the pulse electroplating. This process involves the swift alternating of the potential or current between two different values resulting in a series of pulses of equal amplitude, duration and polarity, separated by zero current. By changing the pulse amplitude and width, it is possible to change the deposited film's composition and thickness.

## Brush Electroplating

A closely related process is brush electroplating, in which localized areas or entire items are plated using a brush saturated with plating solution. The brush, typically a stainless steel body wrapped with a cloth material that both holds the plating solution and prevents direct contact with the item being plated, is connected to the positive side of a low voltage direct-current power source, and the item to be plated connected to the negative. The operator dips the brush in plating solution then applies it to the item, moving the brush continually to get an even distribution of the plating material. Brush electroplating has several advantages over tank plating, including portability, ability to plate items that for some reason cannot be tank plated (one application was the plating of portions of very large decorative support columns in a building restoration), low or no masking requirements, and comparatively low plating solution volume requirements. Disadvantages compared to tank plating can include greater operator involvement (tank plating can frequently be done with minimal attention), and inability to achieve as great a plate thickness.

## Electroless Deposition

Usually an electrolytic cell (consisting of two electrodes, electrolyte, and external source of current) is used for electrodeposition. In contrast, an electroless deposition process uses only one electrode and no external source of electric current. However, the solution for the electroless process needs to contain a reducing agent so that the electrode reaction has the form:

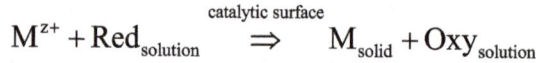

$$M^{z+} + Red_{solution} \quad \overset{\text{catalytic surface}}{\Longrightarrow} \quad M_{solid} + Oxy_{solution}$$

In principle any hydrogen-based reducer can be used although the redox potential of the reducer half-cell must be high enough to overcome the energy barriers inherent in liquid chemistry. Electroless nickel plating uses hypophosphite as the reducer while plating of other metals like silver, gold and copper typically use low molecular weight aldehydes.

A major benefit of this approach over electroplating is that the power sources and plating baths are not needed, reducing the cost of production. The technique can also plate diverse shapes and types of surface. The downside is that the plating process is usually slower and cannot create thick plates of metal. As a consequence of these characteristics, electroless deposition is quite common in the decorative arts.

## Cleanliness

Cleanliness is essential to successful electroplating, since molecular layers of oil can prevent adhesion of the coating. ASTM B322 is a standard guide for cleaning metals prior to electroplating. Cleaning processes include solvent cleaning, hot alkaline detergent cleaning, electro-cleaning, and acid treatment etc. The most common industrial test for cleanliness is the waterbreak test, in which the surface is thoroughly rinsed and held vertical. Hydrophobic contaminants such as oils cause the water to bead and break up, allowing the water to drain rapidly. Perfectly clean metal surfaces are hydrophilic and will retain an unbroken sheet of water that does not bead up or drain off. ASTM F22 describes a version of this test. This test does not detect hydrophilic contaminants, but the electroplating process can displace these easily since the solutions are water-based. Surfactants such as soap reduce the sensitivity of the test and must be thoroughly rinsed off.

## Effects

Electroplating changes the chemical, physical, and mechanical properties of the workpiece. An example of a chemical change is when nickel plating improves corrosion resistance. An example of a physical change is a change in the outward appearance. An example of a mechanical change is a change in tensile strength or surface hardness which is a required attribute in tooling industry. Electroplating of acid gold on underlying copper/nickel-plated circuits reduces contact resistance as well as surface hardness. Copper-plated areas of mild steel act as a mask if case hardening of such areas are not desired. Tin-plated steel is chromium-plated to prevent dulling of the surface due to oxidation of tin.

# History

Luigi Valentino Brugnatelli

Nickel plating

Modern electrochemistry was invented by Italian chemist Luigi Valentino Brugnatelli (it) in 1805. Brugnatelli used his colleague Alessandro Volta's invention of five years earlier, the voltaic pile, to facilitate the first electrodeposition. Brugnatelli's inventions were suppressed by the French Academy of Sciences and did not become used in general industry for the following thirty years. By 1839, scientists in Britain and Russia had independently devised metal deposition processes similar to Brugnatelli's for the copper electroplating of printing press plates.

Boris Jacobi developed electroplating, electrotyping and galvanoplastic sculpture in Russia

Boris Jacobi in Russia not only rediscovered galvanoplastics, but developed electrotyping and galvanoplastic sculpture. Galvanoplastics quickly came into fashion in Russia, with such people as inventor Peter Bagration, scientist Heinrich Lenz and science fiction author Vladimir Odoyevsky all contributing to further development of the technology. Among the most notorious cases of electroplating usage in mid-19th century Russia were gigantic galvanoplastic sculptures of St. Isaac's Cathedral in Saint Petersburg and gold-electroplated dome of the Cathedral of Christ the Saviour in Moscow, the tallest Orthodox church in the world.

Galvanoplastic sculpture on St. Isaac's Cathedral in Saint Petersburg.

The Woolrich Electrical Generator in Thinktank, Birmingham

Soon after, John Wright of Birmingham, England discovered that potassium cyanide was a suitable electrolyte for gold and silver electroplating. Wright's associates, George Elkington and Henry Elkington were awarded the first patents for electroplating in 1840. These two then founded the electroplating industry in Birmingham from where it spread around the world. The Woolrich Electrical Generator of 1844, now in Thinktank, Birmingham Science Museum, is the earliest electrical generator used in an industrial process. It was used by Elkingtons.

The Norddeutsche Affinerie in Hamburg was the first modern electroplating plant starting its production in 1876.

As the science of electrochemistry grew, its relationship to the electroplating process became understood and other types of non-decorative metal electroplating processes were developed. Commercial electroplating of nickel, brass, tin, and zinc were developed by the 1850s. Electroplating baths and equipment based on the patents of the Elkingtons were scaled up to accommodate the plating of numerous large scale objects and for specific manufacturing and engineering applications.

The plating industry received a big boost with the advent of the development of electric generators in the late 19th century. With the higher currents available, metal machine components, hardware, and automotive parts requiring corrosion protection and enhanced wear properties, along with better appearance, could be processed in bulk.

The two World Wars and the growing aviation industry gave impetus to further developments and refinements including such processes as hard chromium plating, bronze alloy plating, sulfamate nickel plating, along with numerous other plating processes. Plating equipment evolved from manually operated tar-lined wooden tanks to automated equipment, capable of processing thousands of kilograms per hour of parts.

One of the American physicist Richard Feynman's first projects was to develop technology for electroplating metal onto plastic. Feynman developed the original idea of his friend into a successful invention, allowing his employer (and friend) to keep commercial promises he had made but could not have fulfilled otherwise.

## Uses

Electroplating is widely used in various industries for coating metal objects with a thin layer of a different metal. The layer of metal deposited has some desired property, which the metal of the object lacks. For example, chromium plating is done on many objects such as car parts, bath taps, kitchen gas burners, wheel rims and many others for the fact that chromium is very corrosion resistant, and thus prolongs the life of the parts. Electroplating has wide usage in industries. It is also used in making inexpensive jewelry. Electroplating increases life of metal and prevents corrosion.

A zinc solution tested in a hull cell

## Hull Cell

The *Hull cell* is a type of test cell used to qualitatively check the condition of an electro-plating bath. It allows for optimization for current density range, optimization of additive concentration, recognition of impurity effects and indication of macro-throwing power capability. The Hull cell replicates the plating bath on a lab scale. It is filled with a sample of the plating solution, an appropriate anode which is connected to a rectifier. The "work" is replaced with a hull cell test panel that will be plated to show the "health" of the bath.

The Hull cell is a trapezoidal container that holds 267 ml of solution. This shape allows one to place the test panel on an angle to the anode. As a result, the deposit is plated at different current densities which can be measured with a hull cell ruler. The solution volume allows for a quantitative optimization of additive concentration: 1 gram addition to 267 mL is equivalent to 0.5 oz/gal in the plating tank.

## Haring-Blum Cell

The Haring-Blum Cell is used to determine the macro throwing power of a plating bath. The cell consists of 2 parallel cathodes with a fixed anode in the middle. The cathodes are at distances from the anode in the ratio of 1:5. The macro throwing power is calculating from the thickness of plating at the two cathodes when a dc current is passed for a specific period of time. The cell is fabricated out of Perspex or glass.

## Nickel Electroplating

Nickel electroplating is a technique of electroplating a thin layer of nickel onto a metal object. The nickel layer can be decorative, provide corrosion resistance, wear resistance, or used to build up worn or undersized parts for salvage purposes.

## Overview

Nickel electroplating is a process of depositing nickel on a metal part. Parts to be plated must be clean and free of dirt, corrosion, and defects before plating can begin. To clean and protect the part during the plating process a combination of heat treating, cleaning, masking, pickling, and etching may be used. Once the piece has been prepared it is immersed into an electrolyte solution and is used as the cathode. The nickel anode is dissolved into the electrolyte to form nickel ions. The ions travel through the solution and deposit on the cathode.

## Types and Chemistry

Watts nickel plating baths can deposit both bright and semi-bright nickel. Bright nickel is typically used for decorative purposes and corrosion protection. Semi-bright deposits are used for engineering nickel where a high luster is not desired.

## Bath Composition

| Chemical Name | Formula | Bright | | Semi-bright | |
|---|---|---|---|---|---|
| | | Metric | US | Metric | US |
| Nickel sulfate | $NiSO_4 \cdot 6H_2O$ | 150–300 g/L | 20–40 oz/gal | 225–300 g/L | 30–40 oz/gal |
| Nickel chloride | $NiCl_2 \cdot 6H_2O$ | 60–150 g/L | 8–20 oz/gal | 30–45 g/L | 4–6 oz/gal |
| Boric acid | $H_3BO_3$ | 37–52 g/L | 5–7 oz/gal | 37–52 g/L | 5–7 oz/gal |

## Operating Conditions

- Temperature: 40-65°C

- Cathode current density: 2-10 A/dm²

- pH: 4.7-5.1

## Brighteners

- Carrier brighteners (e.g. paratoluene sulfonamide, benzene sulphonic acid) in concentration 0.75-23 g/l. Carrier brighteners contain sulfur providing uniform fine Grain structure of the nickel plating.

- Levelers, second class brighteners (e.g. allyl sulfonic acid, formaldehyde chloral hydrate) in concentration 0.0045-0.15 g/l produce (in combination with carrier brighteners) brilliant deposit.

- Auxiliary brighteners (e.g. sodium allyl sulfonate, pyridinum propyl sulfonate) in concentration 0.075-3.8 g/l.

- Inorganic brighteners (e.g. cobalt, zinc) in concentration 0.075-3.8 g/l. Inorganic brighteners impart additional luster to the coating.

Type of the added brighteners and their concentrations determine the deposit appearance: brilliant, bright, semi-bright, satin.

## Nickel Sulfamate

Sulfamate nickel plating is used for many engineering applications. It is deposited for dimensional corrections, abrasion and wear resistance, and corrosion protection. It is also used as an undercoat for chromium.

## Bath Composition

| Chemical name | Formula | Bath concentration | |
|---|---|---|---|
| | | Metric | US |
| Nickel sulfamate | $Ni(SO_3NH_2)_2$ | 300-450 g/l | 40–60 oz/gal |

| Nickel chloride | $NiCl_2 \cdot 6H_2O$ | 0-30 g/l | 0–4 oz/gal |
|---|---|---|---|
| Boric acid | $H_3BO_3$ | 30-45 g/l | 4–6 oz/gal |

## Operating Conditions

- Temperature: 40-60°C

- Cathode current density: 2-25 A/dm²

- pH: 3.5-4.5

## All-chloride

All-Chloride solutions allow for the deposition of thick nickel coatings. They do this because they run at low voltages. However, the deposition has high internal stresses.

| Chemical name | Formula | Bath concentration |
|---|---|---|
| Nickel chloride | $NiCl_2 \cdot 6H_2O$ | 30–40 oz/gal |
| Boric acid | $H_3BO_3$ | 4–4.7 oz/gal |

## Sulfate-chloride

A Sulfate-Chloride bath operates at lower voltages than a Watts bath and provide a higher rate of deposition. Although internal stresses are higher than the Watts bath they are lower than that of an all-chloride bath.

| Chemical name | Formula | Bath concentration |
|---|---|---|
| Nickel sulfate | $NiSO_4 \cdot 6H_2O$ | 20–30 oz/gal |
| Nickel chloride | $NiCl_2 \cdot 6H_2O$ | 20–30 oz/gal |
| Boric acid | $H_3BO_3$ | 4–6 oz/gal |

## All-sulfate

An all-sulfate solution is used for electro-depositing nickel where the anodes are insoluble. For example, plating the insides of steel pipes and fittings may require an anode.

| Chemical name | Formula | Bath concentration |
|---|---|---|
| Nickel sulfate | $NiSO_4 \cdot 6H_2O$ | 30–53 oz/gal |
| Boric acid | $H_3BO_3$ | 4–6 oz/gal |

# Hard Nickel

A hard nickel solution is used when a high tensile strength and hardness deposit is required.

| Chemical name | Formula | Bath concentration | Metric |
|---|---|---|---|
| Nickel sulfate | $NiSO_4 \cdot 6H_2O$ | 24 oz/gal | 179.7g/L |
| Ammonium chloride | $NH_4Cl$ | 3.3 oz/gal | 24.7 g/L |
| Boric acid | $H_3BO_3$ | 4 oz/gal | 29.96 g/L |

# Black Nickel

Black nickel plating is typically plated on brass, bronze, or steel in order to produce a non-reflective surface. This type of plating is used for decorative purposes and does not offer much protection.

| Chemical name | Formula | Bath concentration |
|---|---|---|
| Nickel ammonium sulfate | $NiSO_4 \cdot (NH_4)2SO_4 \cdot 6H_2O$ | 8 oz/gal |
| Zinc sulfate | $ZnSO_4$ | 1.0 oz/gal |
| Sodium thiocyanate | NaCNS | 2 oz/gal |

# Applications

## Decorative Coating

Decorative bright nickel is used in a wide range of applications. It offers a high luster finish, corrosion protection, and wear resistance. In the automotive industry bright nickel can be found on bumpers, rims, exhaust pipes and trim. It is also used for bright work on bicycles and motorcycles. Other applications include hand tools and household items such as lighting and plumbing fixtures, wire racks, firearms, and appliances.

## Engineering Applications

Engineering nickel is used where brightness is not desired. Non decorative applications provide wear and corrosion protection as well as low-stress buildups for dimensional recovery. The method can be used for making nanocomposite wear resistance coatings.

## Electroless Nickel Plating

Electroless nickel plating (EN) is an auto-catalytic chemical technique used to deposit a layer of nickel-phosphorus or nickel-boron alloy on a solid workpiece, such as metal or plastic. The process relies on the presence of a reducing agent, for example hydrated sodium hypophosphite ($NaPO_2H_2 \cdot H_2O$) which reacts with the metal ions to deposit metal. The alloys with different percentage of phosphorus, ranging from 2-5 (low phosphorus) to up to 11-14 (high phosphorus) are possible. The metallurgical properties of alloys depend on the percentage of phosphorus.

## Overview

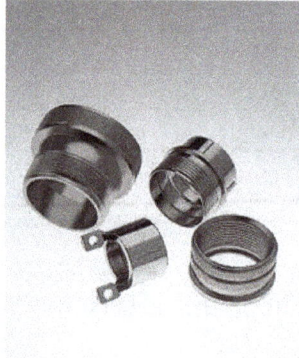

Electroless nickel-plated parts

Electroless nickel plating is an auto-catalytic reaction used to deposit a coating of nickel on a substrate. Unlike electroplating, it is not necessary to pass an electric current through the solution to form a deposit. This plating technique is to prevent corrosion and wear. EN techniques can also be used to manufacture composite coatings by suspending powder in the bath. Electroless nickel plating has several advantages versus electroplating. Free from flux-density and power supply issues, it provides an even deposit regardless of workpiece geometry, and with the proper pre-plate catalyst, can deposit on non-conductive surfaces.

## Historical Overview

The EN plating of metallic nickel from aqueous solution in the presence of hypophosphite was first noted as a chemical accident by Wurtz in 1844. In 1911, Roux reported that metal was inevitably precipitated in the powder form; however these works were not in practical applications. In its early stage, progress in the field remained slow until World War II. In 1946, Brenner and Riddell developed a process for plating the inner walls of tubes with nickel-tungsten alloy, derived from the citrate based bath using an insoluble anode, which brought out the unusual reducing properties of hypophosphite. The U.S. Patent Office says that the patent it issued in 1950 differed from the earlier patent in that Roux reaction was spontaneous and complete, while the Brenner and Riddell process was a controlled catalytic process so that deposition occurred only on catalytic surfaces immersed in the bath. Brenner later wrote that his patent was an accidental discovery similar to the work of Wurtz and Roux, but said that he took out a patent to protect U.S. government rights. In fact, a declassified U.S. Army technical report written in 1963 goes on extensively about Wurtz and Roux work, and gives more of the discovery credit to them than to Brenner. This plating process was attributed to the action of chemical reduction of Ni ions. During the 1954-59 period, Gutzeit at GATC (General American Transportation Corporation) worked on full scale development of electroless plating by chemical reduction alone, as an alternate process to conventional electroplating. Initially, the co-deposition of particles was carried out for electrodepositing Ni-Cr by Odekerken, during the year of 1966. In that study, in an intermediate layer, finely powdered particles like aluminum oxide, polyvinyl chloride (PVC) resin were distributed within a metallic matrix. A layer in the coating is composite but other parts of the coating are not. The first commercial application of their work used the electroless Ni-SiC coatings on the wankel internal combustion engine and another commercial composite incorporating polytetrafluoroethylene (Ni-P-PTFE) was co-deposited, during the year of 1981. However, the co-deposition of diamond and PTFE particles was more difficult than that of composites incorporating $Al_2O_3$ or SiC. The feasibility to incorporate the fine second phase particles, in submicron to nano size, within a metal/alloy matrix has initiated a new generation of composite coatings.

## Pretreatment

Before performing electroless nickel plating, the material to be plated must be cleaned by a series of chemicals; this is known as the pre-treatment process. Failure to remove unwanted "soils" from the part's surface results in poor plating. Each pre-treatment chemical must be followed by water rinsing (normally two to three times) to remove chemicals that may adhere to the surface. De-greasing removes oils from surfaces, whereas acid cleaning removes scaling.

Activation is done with a weak acid etch, or nickel strike or, in the case of non-metallic substrate, a proprietary solution. After the plating process, plated materials must be finished with an anti-oxidation or anti-tarnish chemical such as trisodium phosphate or chromate, followed by water rinsing to prevent staining. The rinse object must then be completely dried or baked to obtain the full hardness of the plating film.

The pre-treatment required for the deposition of nickel and cobalt on a non-conductive surface usually consists of an initial surface preparation to render the substrate hydrophilic. Following this initial step, the surface is activated by a solution of a noble metal, e.g., palladium chloride. Silver nitrate is also used for activating ABS and other plastics. Electroless bath formation varies with the activator. The substrate is now ready for nickel deposition.

## Advantages and Disadvantages

Advantages include:

1. Does not use electrical power.

2. Even coating on parts surface can be achieved.

3. No sophisticated jigs or racks are required.

4.  There is flexibility in plating volume and thickness.

5.  The process can plate recesses and blind holes with stable thickness.

6.  Chemical replenishment can be monitored automatically.

7.  Complex filtration method is not required

8.  Matte, semi-bright or bright finishes can be obtained.

Disadvantages include:

1.  Lifespan of chemicals is limited.

2.  Waste treatment cost is high due to the speedy chemical renewal.

3.  Porous nature of electroless plating leads to inferior material structure compared to electrolytic processes.

Each type of electroless nickel also has particular advantages depending on the application and type of nickel alloy.

## Types

### Low-phosphorus Electroless Nickel

Low-phosphorus treatment is applied for deposits with hardness up to 60 Rockwell C. This type offers a very uniform thickness inside complex configurations as well as outside, which often eliminates grinding after plating. It is also excellent for corrosion resistance in alkaline environments.

### Medium-phosphorus Electroless Nickel

Medium-phosphorus electroless nickel(MPEN) is referred to the nickel-phosphorus alloy deposited by electroless/autocatalytic process in which the resulting alloy consists of medium levels of phosphorus; the definition of medium levels is different in sources of different branches of technology(decorative, industrial, ...). the range accepted as medium levels can be (percent by weight) 4-7 (decorative purpose), 6-9 (industrial sources), or 4-10 (Electronic applications). The EN bath is typically composed of (a.) Nickel source [nickel sulfate ], (b.) Reducing agent [sodium hypophosphite], (c.) Complexing agent; which are necessary to increase phosphite solubility and also to slow down the reaction speed in order to prevent the white-out phenomena but are not co-deposited into the resulting alloy [carboxylic acids or amines], (d.) Stabilizers; which slow down the reduction by co-deposition with the nickel [lead or sulphur or organics], (e.)Buffers [most complexing agents perform double duty as buffers], (f.) Brighteners; mostly co-deposited with nickel and usually can double as stabilizers [cadmium or certain organics], (g.) Surfactants; which reduce surface tension in order reduce pitting and staining [SLS or almost any other surfactant] and (h.) Accelerators; which are added overcome the slow plating rate imparted by complexing agents and usually are co-deposited and can cause discoloration of the deposit [sulfur compounds]. Medium-phosphorus treatment has a high-speed deposit rate and offers bright and semi-bright options for cosmetic particularization. The processing is very stable, used often for Slurry Disposal Industries. This is the most common type of electroless nickel applied.

## High-phosphorus Electroless Nickel

High-phosphorus electroless nickel offers high corrosion resistance, making it ideal for industry standards requiring protection from highly corrosive acidic environments such as oil drilling and coal mining. With microhardness ranging up to 600 VPN, this type ensures very little surface porosity where pit-free plating is required and is not prone to staining. Deposits are non-magnetic when phosphorus content is greater than 11.2%.

## Applications

The most common form of electroless nickel plating produces a nickel phosphorus alloy coating. The phosphorus content in electroless nickel coatings can range from 2% to 13%. It is commonly used in engineering coating applications where wear resistance, hardness and corrosion protection are required. Applications include oil field valves, rotors, drive shafts, paper handling equipment, fuel rails, optical surfaces for diamond turning, door knobs, kitchen utensils, bathroom fixtures, electrical/mechanical tools and office equipment. It is also commonly used as a coating in electronics printed circuit board manufacturing, typically with an overlay of gold to prevent corrosion. This process is known as electroless nickel immersion gold.

Due to the high hardness of the coating it can be used to salvage worn parts. Coatings of 25 to 100 micrometres can be applied and machined back to final dimensions. Its uniform deposition profile mean it can be applied to complex components not readily suited to other hard-wearing coatings like hard chromium.

It is also used extensively in the manufacture of hard disk drives, as a way of providing an atomically smooth coating to the aluminium disks, the magnetic layers are then deposited on top of this film, usually by sputtering and finishing with protective carbon and lubrication layers; these final two layers protect the underlying magnetic layer (media layer) from damage should the read / write head lose its cushion of air and contact the surface.

Its use in the automotive industry for wear resistance has increased significantly. However, it is important to recognise that only End of Life Vehicles Directive or RoHS compliant process types (free from heavy metal stabilizers) may be used for these applications.

## Standards

- AMS-2404

- AMS-C-26074

- ASTM B-733

- ASTM-B-656

- MIL-DTL-32119

## Chrome Plating

Chrome plating (less commonly chromium plating), often referred to simply as chrome, is a technique of electroplating a thin layer of chromium onto a metal object. The chromed layer can be decorative, provide corrosion resistance, ease cleaning procedures, or increase surface hardness. Sometimes a less expensive imitator of chrome may be used for aesthetic purposes.

Decorative chrome plating on a motorcycle

## Process

Chrome plating a component typically includes these stages:

- Degreasing to remove heavy soiling

- Manual cleaning to remove all residual traces of dirt and surface impurities

- Various pretreatments depending on the substrate

- Placement into the chrome plating vat, where it is allowed to warm to solution temperature

- Application of plating current for the required time to attain the desired thickness

There are many variations to this process, depending on the type of substrate being plated. Different substrates need different etching solutions, such as hydrochloric, hydrofluoric, and sulfuric acids. Ferric chloride is also popular for the etching of nimonic alloys. Sometimes the component enters the chrome plating vat while electrically live. Sometimes the component has a conforming anode made from lead/tin or platinized titanium. A typical hard chrome vat plates at about 1 mil (25 µm) per hour.

Various linishing and buffing processes are used in preparing components for decorative chrome plating. The chrome plating chemicals are very toxic. Disposal of chemicals is regulated in most countries.

Some common industry specifications governing the chrome plating process are AMS 2460, AMS 2406, and MIL-STD-1501.

## Hexavalent Chromium

*Hexavalent chromium plating*, also known as *hex-chrome, $Cr^{+6}$,* and *chrome (VI)* plating, uses chromium trioxide (also known as chromic anhydride) as the main ingredient. Hexavalent chromium plating solution is used for decorative and hard plating, along with bright dipping of copper alloys, chromic acid anodizing, and chromate conversion coating.

A typical hexavalent chromium plating process is: (1) activation bath, (2) chromium bath, (3) rinse, and (4) rinse. The activation bath is typically a tank of chromic acid with a reverse current run through it. This etches the work-piece surface and removes any scale. In some cases the activation step is done in the chromium bath. The chromium bath is a mixture of chromium trioxide ($CrO_3$) and sulfuric acid (sulfate, $SO_4$), the ratio of which varies greatly between 75:1 to 250:1 by weight. This results in an extremely acidic bath (pH 0). The temperature and current density in the bath affect the brightness and final coverage. For decorative coating the temperature ranges from 35 to 45 °C (100 to 110 °F), but for hard coating it ranges from 50 to 65 °C (120 to 150 °F). Temperature is also dependent on the current density, because a higher current density requires a higher temperature. Finally, the whole bath is agitated to keep the temperature steady and achieve a uniform deposition.

## Disadvantages

One functional disadvantage of hexavalent chromium plating is low cathode efficiency, which results in bad throwing power. This means it leaves a non-uniform coating, with more on edges and less in inside corners and holes. To overcome this problem the part may be over-plated and ground to size, or auxiliary anodes may be used around the hard-to-plate areas.

From a health standpoint, hexavalent chromium is the most toxic form of chromium. In the U.S. the Environmental Protection Agency regulates it heavily. The EPA lists hexavalent chromium as a hazardous air pollutant because it is a human carcinogen, a "priority pollutant" under the Clean Water Act, and a "hazardous constituent" under the Resource Conservation and Recovery Act. Due to its low cathodic efficiency and high solution viscosity a toxic mist of water and hexavalent chromium is released from the bath. Wet scrubbers are used to control these emissions. The discharge from the wet scrubbers is treated to precipitate the chromium from the solution because it cannot remain in the waste water.

Maintaining a bath surface tension less than 35 dynes/cm requires a frequent cycle of treating the bath with a wetting agent and confirming the effect on surface tension. Traditionally, surface tension is measured with a stalagmometer. This method is, however, tedious and suffers from inaccuracy (errors up to 22 dynes/cm have been reported), and is dependent on the user's experience and capabilities.

Additional toxic waste created from hexavalent chromium baths include lead chromates, which form in the bath because lead anodes are used. Barium is also used to control the sulfate concentration, which leads to the formation of barium sulfate ($BaSO_4$), a hazardous waste.

## Trivalent Chromium

*Trivalent chromium plating*, also known as *tri-chrome*, *Cr$^{+3}$*, and *chrome (III)* plating, uses chromium sulfate or chromium chloride as the main ingredient. Trivalent chromium plating is an alternative to hexavalent chromium in certain applications and thicknesses (e.g. decorative plating).

A trivalent chromium plating process is similar to the hexavalent chromium plating process, except for the bath chemistry and anode composition. There are three main types of trivalent chromium bath configurations:

- A chloride- or sulfate-based electrolyte bath using graphite or composite anodes, plus additives to prevent the oxidation of trivalent chromium to the anodes.

- A sulfate-based bath that uses lead anodes surrounded by boxes filled with sulfuric acid (known as shielded anodes), which keeps the trivalent chromium from oxidizing at the anodes.

- A sulfate-based bath that uses insoluble catalytic anodes, which maintains an electrode potential that prevents oxidation.

The trivalent chromium-plating process can plate the workpieces at a similar temperature, rate and hardness, as compared to hexavalent chromium. Plating thickness ranges from 0.005 to 0.05 mils (0.13 to 1.27 µm).

## Advantages and Disadvantages

The functional advantages of trivalent chromium are higher cathode efficiency and better throwing power. Better throwing power means better production rates. Less energy is required because of the lower current densities required. The process is more robust than hexavalent chromium because it can withstand current interruptions.

From a health standpoint trivalent chromium is intrinsically less toxic than hexavalent chromium. Because of the lower toxicity it is not regulated as strictly, which reduces overhead costs. Other health advantages include higher cathode efficiencies, which lead to less chromium air emissions; lower concentration levels, resulting in less chromium waste and anodes that do not decompose.

One of the disadvantages when the process was first introduced was that decorative customers disapproved of the color differences. Companies now use additives to adjust the color. In hard coating applications, the corrosion resistance of thicker coatings is not quite as good as it is with hexavalent chromium. The cost of the chemicals is greater, but this is usually offset by greater production rates and lower overhead costs. In general, the process must be controlled more closely than in hexavalent chromium plating, especially with respect to metallic impurities. This means processes that are hard to control, such as barrel plating, are much more difficult using a trivalent chromium bath.

## Types

## Decorative

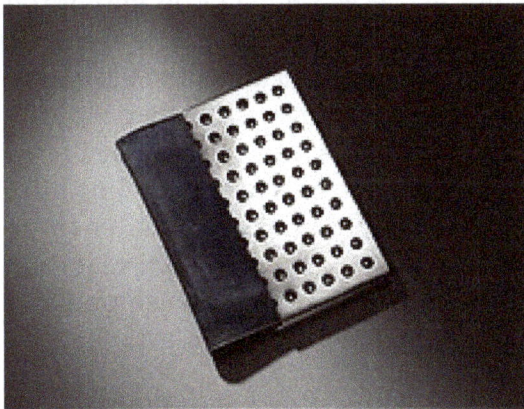

Art Deco portfolio with chrome-plated cover, ca 1925

*Decorative chrome* is designed to be aesthetically pleasing and durable. Thicknesses range from 0.002 to 0.02 mils (0.05 to 0.5 µm), however they are usually between 0.005 and 0.01 mils (0.13 and 0.25 µm). The chromium plating is usually applied over bright nickel plating. Typical base materials include steel, aluminium, plastic, copper alloys, and zinc alloys. Decorative chrome plating is also very corrosion resistant and is often used on car parts, tools and kitchen utensils.

## Hard

Hard chrome plating

*Hard chrome*, also known as *industrial chrome* or *engineered chrome*, is used to reduce friction, improve durability through abrasion tolerance and wear resistance in general, minimize galling or seizing of parts, expand chemical inertness to include a broader set of conditions (especially oxidation resistance, arguably its most famous quality), and bulking material for worn parts to restore their original dimensions. It is very hard, measuring between 65 and 69 HRC (also based on the base metal's hardness). Hard chrome tends to be thicker than decorative chrome, with standard thicknesses in nonsalvage applications ranging from 0.2 to 0.6 mm (200 to 600 µm), but it can be an order of magnitude thicker for extreme wear resistance requirements, in such cases 1 mm (1,000 µm) or thicker provides optimal results. Unfortunately, such thicknesses emphasize the limitations of the process, which are overcome by plating extra thickness then grinding down and lapping to meet requirements or to improve the overall aesthetics of the "chromed" piece. Increasing plating thickness amplifies surface defects and roughness in proportional severity, because hard chrome does not have a leveling effect. Pieces that are not ideally shaped in reference to electric field geometries (nearly every piece sent in for plating, except spheres and egg shaped objects) require even thicker plating to compensate for non-uniform deposition, and much of it is wasted when grinding the piece back to desired dimensions.

Modern "engineered coatings" do not suffer such drawbacks, which often price hard chrome out due to labor costs alone. Hard chrome replacement technologies outperform hard chrome in wear resistance, corrosion resistance, and cost. Rockwell hardness 80 is not extraordinary for such materials. Using spray deposition, uniform thickness that often requires no further polishing or machining is a standard feature of modern engineered coatings. These coatings are often composites of polymers, metals, and ceramic powders or fibers as proprietary embodiments protected by patents or as trade secrets, and thus are usually known by brand names.

Hard chromium plating is subject to different types of quality requirements depending on the application; for instance, the plating on hydraulic piston rods are tested for corrosion resistance with a salt spray test.

## Automotive use

Most bright decorative items affixed to cars are referred to as "chrome," meaning steel that has undergone several plating processes to endure the temperature changes and weather that a car is subject to outdoors. Triple plating is the most expensive and durable process, which involves plating the steel first with copper and then nickel before the chromium plating is applied.

Prior to the application of chrome in the 1920s, nickel electroplating was used. In the short production run prior to the US entry into the Second World War, the government banned plating to save chromium and automobile manufacturers painted the decorative pieces in a complementary color. In the last years of the Korean War, the US contemplated banning chrome in favor of several cheaper processes (such as plating with zinc and then coating with shiny plastic).

In 2007, a Restriction of Hazardous Substances Directive (RoHS) was issued banning several toxic substances for use in the automotive industry in Europe, including hexavalent chromium, which is used in chrome plating. However, chrome plating is metal and contains no hexavalent chromium after it is rinsed, so chrome plating is not banned.

## References

- Todd, Robert H.; Dell K. Allen; Leo Alting (1994). "Surface Coating". Manufacturing Processes Reference Guide. Industrial Press Inc. pp. 454–458. ISBN 0-8311-3049-0

- Lechtman, H. (2014). "A Pre-Columbian Technique for Electrochemical Replacement Plating of Gold and Silver on Copper Objects". JOM. 31 (12): 154. doi:10.1007/BF03354479

- Kuo, Hong-Shi; Hwang, Ing-Shouh; Fu, Tsu-Yi; Lin, Yu-Chun; Chang, Che-Cheng; Tsong, Tien T. (7 November 2006). "Noble Metal/W(111) Single-Atom Tips and Their Field Electron and Ion Emission Characteristics". Japanese Journal of Applied Physics. 45 (11): 8972–8983. doi:10.1143/JJAP.45.8972

- Thomas, John Meurig (1991). Michael Faraday and the Royal Institution: The Genius of Man and Place. Bristol: Hilger. p. 51. ISBN 0750301457

- Pushpavanam, M; Raman, V; Shenoi, B (1981). "Rhodium — Electrodeposition and applications". Surface Technology. 12 (4): 351. doi:10.1016/0376-4583(81)90029-7

- Davis, Joseph R. Nickel, Cobalt, and Their Alloys. ASM International. ISBN 9780871706850. Retrieved 9 August 2016

- Bard, Allan; Inzelt, Gyorgy; Scholz, Fritz, eds. (2012). "Haring-Blum Cell". Electrochemical Dictionary. Springer. p. 444. ISBN 978-3-642-29551-5. doi:10.1007/978-3-642-29551-5_8

- Gal-Or, L.; Silberman, I.; Chaim, R. (1991). "Electrolytic $ZrO_2$ Coatings: I. Electrochemical Aspects". J. Electrochem. Soc. 138 (7): 1939. doi:10.1149/1.2085904

- Wendt, Hartmut; Kreyse, Gerhard (1999). Electrochemical Engineering: Science and Technology in Chemical and Other Industries. Springer. p. 122. ISBN 3540643869

- Stelter, M.; Bombach, H. (2004). "Process Optimization in Copper Electrorefining". Advanced Engineering Materials. 6 (7): 558. doi:10.1002/adem.200400403

# Permissions

All chapters in this book are published with permission under the Creative Commons Attribution Share Alike License or equivalent. Every chapter published in this book has been scrutinized by our experts. Their significance has been extensively debated. The topics covered herein carry significant information for a comprehensive understanding. They may even be implemented as practical applications or may be referred to as a beginning point for further studies.

We would like to thank the editorial team for lending their expertise to make the book truly unique. They have played a crucial role in the development of this book. Without their invaluable contributions this book wouldn't have been possible. They have made vital efforts to compile up to date information on the varied aspects of this subject to make this book a valuable addition to the collection of many professionals and students.

This book was conceptualized with the vision of imparting up-to-date and integrated information in this field. To ensure the same, a matchless editorial board was set up. Every individual on the board went through rigorous rounds of assessment to prove their worth. After which they invested a large part of their time researching and compiling the most relevant data for our readers.

The editorial board has been involved in producing this book since its inception. They have spent rigorous hours researching and exploring the diverse topics which have resulted in the successful publishing of this book. They have passed on their knowledge of decades through this book. To expedite this challenging task, the publisher supported the team at every step. A small team of assistant editors was also appointed to further simplify the editing procedure and attain best results for the readers.

Apart from the editorial board, the designing team has also invested a significant amount of their time in understanding the subject and creating the most relevant covers. They scrutinized every image to scout for the most suitable representation of the subject and create an appropriate cover for the book.

The publishing team has been an ardent support to the editorial, designing and production team. Their endless efforts to recruit the best for this project, has resulted in the accomplishment of this book. They are a veteran in the field of academics and their pool of knowledge is as vast as their experience in printing. Their expertise and guidance has proved useful at every step. Their uncompromising quality standards have made this book an exceptional effort. Their encouragement from time to time has been an inspiration for everyone.

The publisher and the editorial board hope that this book will prove to be a valuable piece of knowledge for students, practitioners and scholars across the globe.

# Index

www.ingramcontent.com/pod-product-compliance
Lightning Source LLC
Chambersburg PA
CBHW062005190326
41458CB00009B/2971